Alber

Jack Steinberg

©Copyright 2015 WE CANT BE BEAT LLC

Copyright 2015 by Jack Steinberg

Published by WE CANT BE BEAT LLC

therealkrob@gmail.com

Table of Contents

Chapter 1: His Birth and Early Life.............................4

Chapter 2: School in Switzerland................................20

Chapter 3: His Early Career...30

Chapter 5: The Birth of the Innovator.......................42

Chapter 6: Continuing in Berlin with More Theories.
....64

Chapter 7: The Theory Brings Him Fame..................89

Chapter 8: His Theories Summarized......................108

Chapter 9: Retirement from Academic Life............123

Chapter 10: Death and Studies of his Brain............131

Chapter 11: How Einstein is Viewed.........................147

Chapter 12: The Overall Conclusion of Einstein.....202

Chapter 1: His Birth and Early Life.

Clearly we need to begin by going all the way back to his early life to start to get a grip of this amazing human being.

Albert Einstein was born on March 14, 1879 in Ulm, Germany into a secular Jewish family and was the elder brother to Maja who was born in 1881. However, even though he was born in Ulm, his family was forced to move to Munich some six weeks after he was born.

According to various reports from the time, the young Einstein was hardly marked out as being a potential genius. Indeed, some historians believe that he was rather slow at learning how to speak much to the disdain of his mother and father who were believed to have been concerned at how their child was apparently not developing in the normal manner. Little did they realize that their child would evolve into one of the sharpest brains that the world has ever seen.

When it comes to his physicality at this age, then there are suggestions that his head was larger than normal for a boy of his age. This physical difference, along with the way in which he rarely spoke, led to a housekeeper believing that he was perhaps 'retarded' in some manner whereas we now know that the complete opposite is true. However, this point is interesting as it does help to show that he was certainly never viewed as being anything special from an early age.

While his mother, Pauline Koch, remained at home in order to bring up the children and run the household, his father, Hermann Einstein initially worked as a bed salesman until he decided to go into business alongside his brother due to the fact that he was also an engineer by trade. Between the two of them, they had created a business by the name of Elektrotechnische Fabrik J.Einstein & Cie. This company was based in Munich, Germany and its main type of business was focused on the manufacturing of electrical equipment. Indeed, the business itself was rather successful at the beginning with them being responsible for adding electricity to the streets of Munich and it is reasonable to assume that being surrounded by this type of technology may have also had a bearing on the life of Einstein.

The family would nowadays be seen as a comfortable middle class family as the business itself was doing quite well and there were not the same everyday stresses on the family that existed for so many other people at this time. His father's business was relatively successful although the earliest years of the life of Albert Einstein do appear to have involved him moving around with evidence of this coming from the various locations where he was at school, which is a topic that we will cover shortly.

Einstein himself wrote about his early years and he draws particular attention to a moment when he was five years old as being an important turning point that may have then gone on to have had an impact on the rest of his life. That moment involved him seeing a compass for the first time and he describes how he felt completely mystified about the way in which the needle of the compass was able to move without anything touching it. He would go on to state how this puzzlement regarding some kind of invisible force would go on to become a lifelong fascination which would ultimately lead to him dedicating so much of his adult life creating the theories for which he is most famous.

If we jump forward slightly in his life to the age of 12, then we see that he appeared to find God and became very religious, which is not a surprise considering his upbringing and being part of a Jewish family. He is believed to have chanted religious sayings on his way to school as well as composing religious songs, so it does appear that religion was playing an important part in his development at least at this stage in his life.

However, it was a period of his life that was quite short after he was introduced to various science books that appealed to him more even though they completely contradicted his religious beliefs. Indeed, it is known that these science books had such a profound effect on him that it led him to do a complete u-turn with his religious beliefs.

There is no doubt that his time in education did play a role in how Einstein developed and that is why we need to spend some time looking at these formative years to get a better grip of his entire life.

His Early Schooling.

His first school days appear to have been at a Catholic school in Munich, his parents were Jewish although they did not practice their faith, and even at the earliest times he was known for having a real interest in a number of subjects. From the little evidence that we have from this time, it is believed that he had a penchant for mathematics and this was something that would be further developed as he moved through his school career.

When it comes to his early schooling, then we know that he was then enrolled at the Luitpold Gymnasium although this was not the best of times for the young Einstein. It appears to be the case, with Einstein himself confirming this, that he was none best pleased with the way in which the school was run. He was seen as being a loner and was troubled by a speech issue that was not being addressed and even though he is known to have developed a love for things such as classical music he was largely alienated with this school in particular being a rather troubled period in his life.

It is worth mentioning that his love of classical music stemmed from the fact that he was taught how to play the violin from six years of age. This was something that was forced on him by his mother although he was initially not too keen, he did eventually admit that it helped him in later life. For some, this pursuit has been interpreted as playing a role in the development of key areas of his brain that then helped him to formulate ideas and theories later on in life, but that is one area that we will focus on in another chapter.

It also appears that he had an issue with the Prussian style of education that was preferred at this school as there was also little opportunity for him to be able to express himself in a creative manner. To Einstein, it was too strict and focused on the wrong things as he was already developing ideas connected to science, which was something that the school itself appeared to have no desire in pursuing. Indeed, such was his frustration that it appears to have been the case that he then showed this anger at various times leading to one teacher remarking to him that he would never amount to anything in his life due mainly to the way in which he was unable to follow the type of education on offer at the school. This would not be the first time that this allegation would be thrown at him during his education.

His penchant for criticizing the authority of the school, and willingness to describe how he was unhappy, even led to him being given the nickname of 'Beidermeier' by his fellow students. This translates roughly as 'Honest John' and it was purely because of the way in which he had a tendency to just voice his true feelings no matter if it would then get him into trouble. There is no doubt that he was certainly different to the other students and not always in a good way.

The problem for Einstein was the way in which he would simply get into trouble due to his views. Indeed, Einstein himself has described one such incident at school that gives us a glimpse into what was going on in his world at this time.

"When I was in the seventh grade at the Lutpold Gymnasium, I was summoned by my home-room teacher, who expressed the wish that I leave the school. To my remark that I had done nothing wrong, he replied only 'Your mere presence spoils the respect of the class for me'."

In other words, Einstein had ruffled a few feathers with his actions although many do see it as being rather shocking that his teacher would voice such an opinion.

However, even though he was clearly frustrated at the lack of opportunities at the Luitpold Gymnasium, it did not prevent him from showing a real interest in the world of mathematics. It is known that even from an early age he was showing a talent for this subject that was beyond his relatively young age and was creating mathematical models that were more advanced than others around him. This particular interest in mathematics was something that would only develop even further over the next few years.

We know that from approximately 1891 that Einstein had successfully managed to teach himself Euclidean geometry merely from a school book and he then switched his attention to calculus, which would go on to serve him well later on in life. It is believed that through his love of geometry that he then began to really understand the power of deductive reasoning with this allowing him to begin to think in a broader manner about so many aspects of life.

The Influence of Max Talmud.

Clearly the early teenage years were tough for Einstein since he was having issues with his schooling and life at home was becoming tougher due to his father's business coming under intense pressure. However, we do need to go back in time just a few years to the late 1880's in order to talk about an individual that would go on to become a major influence in the life of Albert Einstein, a Polish individual by the name of Max Talmud.

Max Talmud was a medical student who had become friends with the Einstein family and would occasionally come to their home for dinner. Both Talmud and Einstein struck up a friendship with Talmud becoming something akin to an informal tutor and was responsible for helping to really open up the mind of the young Einstein to a world outside of religion. Einstein himself would go on to pay homage later in life to the influence that Talmud had on him at this time.

This was a friendship that would develop over the next few years and their discussions are seen as laying the foundations for a number of the inquisitive ideas and concepts that plagued the mind of Einstein for his entire life.

We know that Talmud was responsible for really harnessing the interest in mathematics and science that already existed within the mind of Einstein, but for the first time in his life he was able to really talk to another individual that had the same core interests and was able to answer some of the rather searching questions that were floating around in his mind. However, Talmud also introduced Einstein to the concept of philosophy with this also playing a role in creating various ideas and questions in his mind that would go on to shape the way that he viewed the world and sought answers to questions that would change the way in which we now view everything that exists.

Talmud was also the individual who was responsible for introducing Einstein to a series of science books that had been written by Aaron Bernstein and that one decision would go on to play a pivotal role in the development of the young scientist. By the time that he was 16, Einstein was looking at some of the theories and ideas contained within this series of books from a different perspective. Indeed, in one book the author described a scene whereby he was imagining riding alongside electricity that was being carried within wire. This led to Einstein beginning to create a number of interesting questions within his own mind although his questions focused more on light rather than electricity. This concept of looking into light was something that would go on to largely dominate his way of thinking for the next decade.

To him, he was more interested in what would happen if you could run alongside that beam of light, what would it look like? How would it react? These questions were certainly regarded as being rather strange for a teenager to be considering, but we do know that Einstein had already spent time as a younger child thinking about light in general. Indeed, it is rather peculiar to think that he had already understood that stationary light waves had never been seen at this point and it is fair to say that he then set out to change that particular theory.

To Einstein, the issue was further complicated by the way in which he presumed that if light was indeed a wave, then the light beam itself would be stationary. To a young mind this kind of thought would have been too much to comprehend, but thanks in part to his tutoring by Talmud it does seem that Einstein had a desire to tackle this issue and come up with some kind of answer.

As we said earlier, Einstein was approximately 16 at this time, but that did not stop him producing his very first scientific paper titled 'The Investigation of the State of Aether in Magnetic Fields' although we will look at it in slightly more detail later on.

The Latter Teenage Years.

Growing up was certainly not easy for Albert Einstein and part of the problem was the way in which his father's business would either be struggling or close to failing on a number of occasions. This led to some upheaval within his family and indeed it is known that around the age of 16 his father was forced to move to Milan to take up a job after his company lost an important contract. The problem for his father was that the electrical components that they were dealing with were becoming outdated and overtaken with new technology. They were unable to adapt to the changes meaning that they lost out on major contracts forcing them into the difficult decision to close their business.

Due to this failure, he was forced to take up a job working with a relative and even though they moved to Milan the young Einstein was put into a boarding school and was left to complete his education all on his own. The problem was that Einstein was already suffering from severe boredom with school life and this feeling was further compounded by the fact that he was edging ever closer to the age whereby he would be forced into joining the army for his national service.

It is also known that Einstein was, at this stage in his life at least, attempting to follow the wishes of his father and to enter the world of electrical engineering. His schooling does tend to back this up since he was clearly excelling at certain key subjects, but in particular mathematics and physics, but he had also developed a belief that the school was, in effect, holding him back. The young Einstein wishes to express himself in a more free-flowing and creative manner, but as we said earlier the structure of the school meant that this was impossible for him to do. In short, he felt that he was being pushed in a completely different direction to which he wanted to go and when you add in his fear of national service, you get an individual who is rather troubled in his own way and one that felt he had to take some kind of action before it was too late.

This thought was clearly too much to bear and it ultimately led to him fleeing school in Munich and somehow making his way to Milan where he suddenly pitched up on the doorstep of his parents house much to their surprise. However, an alternative theory is that a teacher actually suggested he leave school, which was probably in jest or thanks to a sense of frustration about Einstein, with him then taking the suggestion seriously and doing just that.

The problem that his parents faced was that they could only see that their child would be regarded as being a school dropout at a time when this would have meant that his prospects would have been rather poor to say the least. They must have been concerned and partly believed that he would never amount to anything, but there was some luck on his side that would ultimately pave the way for him to become one of the greatest scientific minds that the world has ever seen.

Chapter 2: School in Switzerland.

The lucky break for Einstein came in Switzerland with him being able to enrol in the Swiss Federal Polytechnic School which was based in Zurich. However, it is known that he only gained admission due to his extremely high scores in the entrance exam for the subjects of mathematics and physics. Indeed, he is known to have excelled at these two subjects, but scores in other subjects let him down; however, the school at least saw some promise in those two subjects allowing him to enrol, but with some conditions attached. The problem was that he was unable to simply enrol and begin his studies as he was also required to complete his pre-university education before he would be allowed to study there and as a result it is known that he moved to a High School in Aarau, Switzerland which was run by an individual by the name of Jost Winteler.

In order to attend this school it did mean that Einstein spent time boarding with the Winteler family and this was something that would ultimately change his life. He was known to have had issues with a number of subjects including French and Chemistry as well as Biology, so those were the subjects that he was forced into focusing on during his time at the school. However, he certainly had no need to expand his knowledge of mathematics or physics in which he excelled.

He also became very close friends with the Winteler family and Einstein himself has even hinted that, at one point, he fell in love with the daughter of Jost Winteler by the name of Marie with her becoming his first love. These connections with the Winteler family would remain with Einstein for the rest of his life and even his sister would eventually marry one of the sons of the Winteler family.

We know that he graduated from this school in 1896 and he was then allowed to enter the university after successfully obtaining the correct qualifications. This meant that he was able to follow the path that he himself had predicted when entering high school at Aarau as he is quoted as having written.

"If I were to have the good fortune to pass my examinations, I would go to Zurich. I would stay there for four years in order to study mathematics and physics. I imagine myself becoming a teacher in those branches of the natural science, choosing the theoretical part of them. Here are the reasons which lead me to this plan. Above all, it is my disposition for abstract and mathematical thought, and my lack of imagination and practical ability."

Thanks to his grades from high school, his dream was then able to become a reality. However, we should also mention at this point that he still feared being called up for national service, so in 1896 we see him renouncing his German citizenship in order to avoid this and he actually remained stateless until 1901 when he took up Swiss citizenship. This changing of citizenship would become a recurring theme throughout the life of Einstein although for a variety of reasons.

Moving to Zurich.

The move to Zurich was a life changing event. By this stage, Albert Einstein was still only 17 years of age. He had been successful in avoiding national service due to his move to renounce his German citizenship allowing him to focus on his studies. As an aside, after he gained Swiss citizenship in 1901 he then managed to avoid their national service as he had flat feet as well as varicose veins. He certainly appeared to be a lucky individual in avoiding this kind of service, which was quite apt for such a strong pacifist.

After being accepted into the university in Zurich to study in the department dedicated to producing teachers in the world of mathematics and physics, Einstein would have a relatively quiet four years there with him becoming quite involved in his studies. However, it was also the time where he met who would become his future wife, a Serbian student by the name of Mileva Maric. There was a problem for him in having met her as he realized early on that he had the desire to marry her, but his own parents were against the union and he was too poor to afford the wedding on his own.

One thing that we do know about this relationship is that even though they were not married at first, they did have an illegitimate daughter in 1902 by the name of Lieserl. Sadly, we do not know too much about what happened to her as she does not tend to feature too often in the story of Einstein. Stories from various sources point to her having been put up for adoption in Hungary although there is little proof available to either confirm or deny this.

University Life.

His time at university was certainly an eye opener for him as it allowed him to focus more on the areas that were of interest to him although he was still open to clashing with authority. Indeed, Einstein was often quoted throughout his life that his time in Zurich were amongst the happiest years of his life as it was the time where he was able to build a number of lifelong friendships with like-minded people. This included the likes of Michele Besso who Einstein would have extensive conversations with on subjects such as time and space and how the two were potentially linked. Additionally, he was able to form a relationship with a mathematician by the name of Marcel Grossmann who would also develop into a trusted companion throughout his life.

However, university life was not always plain sailing for him. He appears to have made a habit of not going to lectures as he had a preference of staying at home in order to study more advanced aspects of his subjects on his own and that did result in him making life harder for himself. One major problem was that a number of lecturers then took umbrage against his lack of participation and this is one potential reason why he then found it difficult to get a job after he left the university. After all, he would be required to ask for recommendations from them and they were none best pleased as to the way they had been treated, so they were rather reluctant to provide him with a glowing reference.

We can look at the views of two of his teachers to see how they certainly did not hold him in high regard during his time at the university. Indeed, his maths professor, Hermann Minkowski referred to Einstein as being a 'lazy dog' whereas his physics professor, Heinrich Weber, went even further in his opinion. He had really gone out of his way to help Einstein in his early days at the university, but it is clear that he was becoming exasperated at the way in which Einstein was acting. He is even quoted as saying *'You're a clever fellow, but you have one fault: You won't let anyone tell you a thing!'*

This particular quote is certainly something that would be thrown at Einstein throughout his life due to his lack of being able to entertain certain ideas that directly contradicted his own theories.

Issues with the Course.

It seems that one of the major issues that Einstein had with his time at the university was that he felt disappointed at some of the theories regarding physics that had been left out. He had a desire to study the latest concepts, but he believed that the university was stuck in the past and was, to a certain extent, hindering him. This can be shown by the way in which Weber viewed certain theories as being far too modern to be taught at a university, which was in direct competition to what Einstein believed as he preferred to look at the more recent advances in order to determine if he could take them further.

He appears to have had a particular interest in the theory by James Clerk Maxwell surrounding electromagnetic field, but that was something that the university appeared to have no interest in covering. As a direct result of avoiding lectures due to his disappointment in the course, it did also mean that he was forced into using the notes of other students in order to then sit the examinations. In particular, he relied on Grossmann for help and it certainly seems to be the case that he was still able to pass with flying colors. It has been argued by many that he would have actually failed two of his examinations at the university had it not been for Grossmann. This in itself is pretty amazing to even stop and think about when you consider he would then be viewed as a genius by the time he reached his twenties.

From another perspective, this can perhaps be seen as an example of how Einstein was extremely focused on his own ideas even from a relatively early age. Perhaps it was a sign of confidence in himself although with his struggles after university, this confidence would have surely been dented.

However, there was one aspect that Einstein struggled with and that was in his third year at university and the 'Physics Practical Course of Beginners' by Jean Pernet. This in itself is a surprise as you hardly imagine that Einstein would have ever had any difficulty surrounding something that was connected to being a beginner with physics. It is therefore assumed that something else was going on whereby he lost interest in this part of his course with some coming to the conclusion that he had an issue with the lecturer. Indeed, thanks to his university records we do know that it was around this time that he was issued with a reprimand due to effectively playing truant and, as a result, he was awarded the lowest possible mark as a form of punishment.

Einstein eventually graduated from university in 1900 with a diploma, but if he believed that this would be the start of his career, it would not turn out that way, at least not in the beginning.

Chapter 3: His Early Career.

We have previously alluded to Einstein having a number of problems after graduating from university, but the scale of this problem is difficult for us to grasp due to our understanding of how he then developed throughout his life. It does at least appear to be the case that his inability to listen to the advice of others may have hindered him, at the outset anyway.

It is first of all interesting to note that he wished to become a teacher in the fields of mathematics and physics and out of his entire class he was one of only four that struggled to get a placement. However, we do know that his old lecturers played a role in this and in particular Heinrich Weber who had been particularly annoyed by the way in which Einstein would avoid his lectures on a regular basis.

The problem for Einstein was that he appears to have made a habit of asking Weber for a recommendation although this was clearly the wrong thing to do as

Weber then made a habit of not exactly giving Einstein a glowing reference. Indeed, Einstein himself was even quoted later on in his life as saying:-

"I would have found a job long ago if Weber had not played a dishonest game with me."

In other words, Einstein clearly felt that his potential future career as a teacher was being thwarted by this individual and that in itself is an amazing thing to take into account when you then consider what he went on to achieve in his life. We do know that he spent in the region of two years trying to find a job in teaching but to no avail with him failing to secure even one single job during that time. Indeed, it was not even in the world of teaching that he was unable to get any employment as any job seemed to be out of his reach due to the poor recommendations that came from his university.

To actually show how difficult it was for him at this time we only have to look at the fact that he was turned down for positions at a range of universities across a number of countries in Europe including the Netherlands, Germany, and Italy. It did at least seem as if his name had been sullied at this point making it virtually impossible for him to pursue his own particular dream of becoming a teacher.

At this point it is known that his own father believed that his son would never amount to anything and there was a sense of disappointment surrounding Einstein from both himself as well as his family. This would be seen as among his darkest days and even Einstein himself would have found it impossible to then see what the future would hold.

His First Move into Employment.

Between 1901 and 1903 Einstein was largely struggling and in 1902 he reached what would become his lowest ebb when the business run by his father collapsed and he found himself trying to work as a tutor for children. However, he was apparently even being fired from these positions perhaps because his heart was not in this aspect of teaching. We are also aware of him being given a teaching position, but only for two months, at a Technical College at Winterthur which only really served to give him a taste of the type of life that he had initially wanted to lead.

His fortunes did eventually take a turn for the good in the later part of 1901 thanks to his old university friend Marcel Grossmann. Through this one single contact, he was put forward for a position in Bern, Switzerland where he would work as a clerk in the Swiss Patent Office. This in itself was not the kind of position that he had initially wanted to find himself in, but after several years of feeling like a failure it was clearly something that he wished to take advantage of so was quite happy to take up the position. He began his career at the patent office in the middle of June 1902 and it would go on to be the start of something quite wonderful for him, but not in the way that he had expected. For Einstein, getting this job was important although it did mean that he was ranked the lowest of the low within the office. It took him some time before he could be promoted from the third level to the second level and to then be paid a slightly higher salary. Considering his excellence in certain subjects this was certainly an inauspicious start in life.

This high point of at least getting a job was also matched by a low point in his life as his father became very ill and then died. It is known that Einstein suffered a great deal from his death as he struggled to come to terms with the idea that his father had effectively died believing that his son was a failure. Quite how that then affected the rest of his life is unknown, but there is no doubt that it did depress him a great deal for some time.

The job in Bern proved to be rather fruitful even though he was not being paid a fantastic salary. However, it was still enough of an income for him to then be able to afford marrying his sweetheart, Maric with them being wed on 6th January, 1903. It is perhaps no coincidence that they married after the death of his father considering both families were firmly against them.

It was clear from early on that the job at the patent office was below his capabilities, but it also worked in his favor in other ways. We know that he would complete his work of analyzing the different patent applications in a short period of time and this freed up more time for him to focus on the various day dreams that he had held since a teenager regarding various theories. However, we are not exactly sure how this free time led to his key theories although we can presume that it played an important role considering the release of five papers in one year in 1905.

Of course, Einstein had come a long way since he was 16 and dreaming about light waves and beams and whether or not they were stationary. He had spent time in Zurich studying the work of Maxwell and this, in turn, allowed him to come to a number of conclusions regarding light and various equations and indeed this period is attributed to helping him develop a number of his other theories that he would go on to become famous for throughout his career.

Considering his relatively young age, it is amazing to then discover that Einstein was responsible for noticing a fact that had been overlooked by the esteemed Maxwell. The fact that was overlooked? That the actual speed of light remains constant no matter how fast you yourself may move. To Einstein this was obvious. To others, it was regarded as a major breakthrough that had somehow not been noticed before.

He was helped in this job by the fact that his main area to focus on included patents that dealt with electromagnetic devices. The majority of the patents that he dealt with included answering questions that were based largely around the way in which electrical signals were transmitted as well as issues related to the synchronization of time via electrical and mechanical means. There is no doubt that having to deal with these points on a regular basis did stir up a whole host of thoughts inside the mind of Einstein and it is perhaps no surprise that just a few years later he had what has been classed as a major breakthrough in his career.

Perhaps his career in the office not only helped to shape the mind of Einstein even further, but it helped to shape the future of our own understanding of the world.

Chapter 4: Building a Life.

However, prior to going into details regarding his major breakthrough, we should spend some time looking at another area of his life that was certainly taking shape, that of his ever growing family.

After meeting Maric at university, where she was the only woman on his physics course, the young couple were troubled by the fact that they initially did not have the money available to get married and also the mother of Einstein was none best pleased due to her Serbian heritage. However, just prior to his death, Einstein's father gave his blessing for the two of them to be married opening up the door for them to do so in 1903.

The problem is that we do know that they previously had a daughter in 1902 who was, obviously, illegitimate and there are several theories as to what happened to her. Some historians and scholars state that she died of scarlet fever while others believe that she was put up for adoption although there is a third theory that the family of Maric back in modern-day Serbia took her although she does appear to vanish from history relatively early on.

Even though they had previously had an illegitimate child, the two kept in touch while Einstein was searching for a job via sending letters to one another. It is known that they had a number of illicit trysts during this time in order to help keep their romance alive. A year after marrying, the couple had another child, but this time a son by the name of Hans Albert who was born in Bern in 1904 and he was then followed in 1910 by yet another son called Eduard.

However, there were a number of problems related to his marriage including the fact that thanks to his miraculous year of 1905, Einstein was becoming quite famous and was being asked to give talks at a number of locations and universities around Europe. His fame was increasing and it was also reported that he became completely consumed by his thoughts on issues such as relativity. As a result, he was largely putting his family to one side although there were also arguments surrounding their finances. Considering he had a wife and child to support, working as a lowly patent officer was certainly not enough.

In order to tackle the issue with money, Einstein applied for another job and as a result it led him to move to the University of Berlin, but this in turn made matters worse. Indeed, it took him away from his family for extended periods of time and it is also know that Einstein entered into an affair with his cousin called Elsa Lowenthal and this was something that would continue right up until he was divorced by his wife as he would go on to marry Elsa.

It does seem rather unfortunate that the private life of Albert Einstein was in so much turmoil at the point where his career was showing an upturn in its fortunes. It was becoming clear that Einstein was a man in demand and that was something that would only become more apparent even though his career was largely brought to a halt by the First World War.

We do know that his marriage to Elsa was reasonably happy although there were no further children. It is believed that he fell for her after she nursed him back to health after a serious illness while in Berlin and they would be together for some 20 years until her death in 1936. After that death it is felt that Einstein was so deeply affected that his private life was never the same again.

The one thing that we can say is that the prime of his career, at least from a production of theories point of view, also coincided with having a strong woman beside him. Some scholars even go as far as to state that his first wife had more of an impact on his theories than is often attributed to her since she too was a physicist. After the death of his second wife, he became too consumed by his work and focusing on certain theories that still remain impossible to prove even today, but we will look at those in the future.

Chapter 5: The Birth of the Innovator.

It is easy for us to identify Einstein as being an innovator and he certainly burst onto the physics scene with a bang and in a way that nobody else had ever been able to do before. It was almost as if Einstein was feeling desperate to put his work out there in order to either gauge scientific reaction or to even clear those thoughts from his mind to allow him to focus on other areas.

1905 is widely regarded as being a major breakthrough year for Einstein as it saw him publishing four different papers on a variety of topics. This would have always been regarded as a lot of work, but the fact that he was able to do this while also working at the Patent Office does show how dedicated he was to his chosen craft.

He was of course helped by the fact that he had a tendency to finish his work earlier than anticipated since he was dealing with patents and questions that were directly related to a number of areas that were of real interest to him. This then led to him freeing up time that could then be dedicated to working on his own ideas which, in turn, would result in him paying particular attention to a number of theories that he had been developing in his mind even from his time at university.

We previously stated how Einstein felt drawn towards theories that had been put forward by Maxwell and indeed it appears to have been the case that he focused on them yet again in the years that immediately followed his graduation from university. In particular, he felt drawn towards the idea of light and the work that Maxwell had been involved in regarding the nature of light through a series of equations. However, Einstein would go on to not only deal with those equations, but to take the understanding of them further than ever before by discovering facts that even Maxwell himself had been unable to see when at his peak.

The difficulty for Einstein, at least in his eyes, was that theories put forward by Maxwell appeared to completely contradict those that had been put forward by Newton some time earlier. Einstein had been able to discover something very important early on in his work in that the speed of light remained constant no matter your own pace and that was where the issue with the work by Newton really came into play.

The problem was that this theory held by Einstein went against the laws of motion as published by Newton with this having been the accepted theory in the world of physics for quite some time. However, it would go on to help form the basis of his most famous work, the concept of relativity and that is something that we will of course examine later on. Indeed, this would not be the first time that Einstein would tackle theories as proposed by Newton.

It is important to remember that Einstein was still relatively young at the time that he produced these works and he was certainly not seen as being anybody with any real promise at this moment in time either. However, that was something that was about to change and it would do so thanks to his work in just one year. At this point, he would certainly make the world of physics stand up and take notice.

His First Paper as an Adult.

The first paper that he produced as an adult was focused on work that had previously been carried out by another German physicist by the name of Max Planck. His work had primarily been focused on the quantum theory but, as was the norm, it did lead to Einstein taking the theory further than ever before. Planck was an individual that Einstein treated with a sense of adoration although that did not make him immune from coming under close examination.

Max Planck had previously applied the quantum theory to light with Einstein then looking at the impact that it had on the photoelectric effect which is when some kind of material would emit an electrical charge whenever it was hit by light. When it comes to Einstein, he showed that protons were able to knock out electrons in a precise manner thanks to this effect.

You have to remember that Einstein had been fascinated by light for a number of years and had sought to explain various aspects of its power even as a teenager. However, thanks to this paper he managed to provide the theory that light is capable of interacting with matter via energy, which up until now was something that had only been suggested as a mathematical possibility by Max Planck.

This paper was published, but it did not exactly receive widespread acclaim although it did at least show those in the field that here was a rather interesting individual who was looking at current theories and not only putting them to the test, but expanding on them further. It was something that would occur time and time again throughout 1905 although not everything was focused on the photoelectric effect that was initially covered in this first paper. Also, it made Planck sit up and take notice of Einstein, something that would go on to help him later on in his career.

His Second Paper.

The second paper that was published in 1905 and attributed to Einstein focused more on the world of atoms rather than protons and the impact that they had on electrons. Indeed, this paper was the very first time that the existence of atoms had been proven via experiments which in itself was a major development. His experiment focused on the study of tiny particles that had been suspended in water and their motion with this being referred to as the Browning Motion. By studying them, he was then able to calculate not only the size of the atoms that were fighting against one another, but also an additional feature known as Avogadro's Number.

This paper was able to confirm the atomic theory that had been floating around for some time, but Einstein was able to take things further and show that all matter is indeed constructed of both atoms and molecules. However, he remained unconvinced that atoms could be split even though he did consider it at one point. This was a theory that would be discovered later and by other physicists.

It is amazing to consider that up until this point the existence of atoms had just been a theory and that some of the greats of physics had not managed to successfully prove it until Einstein came along and managed to do so successfully in his spare time. This is, at times, seen as being a paper that really did help others to take notice of what Einstein was capable of doing, but as we can now see by looking back at history he was not finished with his plans for 1905.

His Third Paper.

It was his third paper of the year that was to become his most famous work of this period as it was a paper that led him back to issues surrounding the two greats of physics that had preceded him, Maxwell and Newton. This in itself was no surprise as Einstein had been a fan particularly of Maxwell since his time at university. To him, being able to use the theories as put forward by Maxwell was something that he knew he could trust even when he was trying to expand on the work that had been done previously.

This paper was given the name "On the Electrodynamics of Moving Bodies" and this is often regarded as being the work that propelled him to effective stardom within the world of physics. As a result, we can also perhaps view this work as the initial trigger for the downturn in his marriage since it would lead to him giving talks in a variety of locations that put extra stress on his home life.

The work itself was seen as being rather important as the main aim of Einstein was to tackle the issue of there being a direct contradiction between the work of Newton and Maxwell. These two theories were regarded as being main principles of the world of physics and to have them working against one another was something that Einstein felt could not go on any longer.

The theory held by Newton was focused on the issue of time and space and concepts surrounding the two whereas the theory held by Maxwell was that the speed of light was indeed a constant.

These two theories had troubled Einstein for some time as he had spent countless hours pouring over their work and calculations in order to try to make sense of it all until he had to come to the conclusion that something was wrong and the two theories were unable to stand side by side with one another. Instead, it eventually led to Einstein producing his own special theory of relativity whereby he declared that the actual laws of physics must be a constant and the exact same no matter if the objects are actually moving in different inertial frames. In short, the objects are moving at a constant speed that are all relative to one another.

This paper also went further in that it then suggested that the actual speed of light is also a constant no matter the inertial frame and in producing these two different parts to his theory it is accepted that Einstein opened up a new world for discussion. However, we should point out that his main theories surrounding the concept of relativity was something that he would publish later on in life although it could be argued that this laid the foundation for his future ideas.

His Fourth Paper.

His fourth and final paper of this productive year led to the creation of one of his most famous creations, the equation E=mc2. Little did Einstein realize that this equation would become known around the world and have such a profound effect on science in general.

The entire paper focused on the relationship between mass and energy and this in itself was turning previous theories on their head. Other physicists had believed that they were individual concepts, but Einstein disagreed and worked at proving that there was a correlation between the two. This, to him, was a fundamental part of physics and nature and being given the opportunity to prove it was something of interest to him.

This paper was also the first time in which there was an attempt to actually explain and understand the energy source from the sun although perhaps the most surprising part is that it seems that this paper was nothing more than an afterthought by Einstein. It is difficult to imagine how different things may have been had he decided not to publish this one single equation as it has perhaps turned out to be the most popular thing that is quoted by people around the world regarding the work of Einstein.

Overall, these papers showed that Einstein was capable of tackling some of the biggest questions in the world of physics head on and he did so without displaying any fear for the major names that he was tackling. This year was indeed the one that really managed to propel Einstein to a new level and he finished the year by achieving his doctorate although he would remain in his position at the Patent Office for some time to come.

We should also mention at this point that the paper regarding relativity was not the first time that this theory had been put forward. Indeed, aspects of it had already been suggested by physicists such as Hendrik Lorentz and Henri Poincare, but Einstein was the first individual that was capable of pulling together different aspects of the theory and combining it into something that could be understood.

In actual fact, both Lorentz and Poincare believed that the concept of relativity was nothing more than a type of motion in the ether and they clearly did not place too much emphasis on it, or they were simply incapable of pulling together the theory. However, Einstein was different in that he discovered that it was actually a universal law of nature rather than just being something akin to a sideshow.

The only problem was that Einstein may have believed that this year with the four different papers would have changed the opinion of those within the world of physics. As it turned out, this would not be the case.

Being Ignored.

As we look back at this point in the life of Einstein, and being aware of what he went on to achieve, it is rather surprising for us to discover that his papers in 1905 were largely being ignored by the physics community after they were first published.

Part of the problem was that a number of scientists saw his work as being rather controversial since here was an upstart that was trying to change the way that people had been thinking for some considerable period of time. However, that all changed when Max Planck himself began to give them some credit as here was a well established and respected name stating that the points that were being made in these papers were things that people should pay attention to. To be honest, considering the first paper produced by Einstein focused on his own theory, then you can understand why he took an interest since one of his own theories had been turned from a theory into a reality.

As soon as Planck himself began to take an interest, it seems as if the physics community changed tact as well as changing their opinion of Einstein. He still had to wait while the community went about conducting various experiments to put his theories to the test, but when those experiments confirmed what was in his various papers, then we see the community taking him more seriously as a result.

Once again, it is amazing to think about how different things could have been had Planck not picked up on the papers or realized their significance and took it upon himself to spread the word of the work of Einstein. Would he have slipped backwards and simply remained a physics teacher? Perhaps that really would have been the case and this could only have lead to the world missing out on a wonderful mind.

Thanks to Planck, it led to Einstein being invited to give lectures at a number of different locations and this is the point that we alluded to earlier regarding the gradual failing of his married. At the same time, he was also being subjected to various work offers at a variety of universities around Europe including some rather prestigious locations. Included in this list were institutions such as the University of Zurich, the University of Prague, and also the Swiss Federal Institute of Technology. However, it was the University of Berlin that he would ultimately move to although that was not until a few years after the production of his papers.

Furthering his Teaching Career.

Considering the eventual impact that his papers from 1905 had on the physics world, it is certainly rather surprising to discover that Einstein remained at the Patent Office for several years. However, he eventually left that position in 1908 to take up a teaching post at the University of Bern. Quite what took him that long to change things is something we are not privy to.

This was clearly a major leap forward for Einstein as he had been seeking this kind of employment since he graduated and it is clear that he was more than happy to take up this position in 1908 since he was not required to move away from home. However, his family life was still regarded as being very healthy although this was something that would of course change in a few years. At first, it must have appeared as if everything was going to work out well after several years of struggling by.

This position at the University of Bern did not last for an extended period of time as he then switched to the University of Prague although he stayed there for just one year. After this, he felt the draw back to Zurich with him taking up a teaching position at the university.

The problem for us at this point is that we actually know very little about these few years of his academic life. We do, however, know that he was still working on a number of papers that would be published later on in his life. At the same time, he was clearly doing work behind the scenes on what would become his most famous work, the Theory of General Relativity.

Furthermore, he spent a number of years pouring over one of his own theories that was guilty of almost constantly annoying him due to him identifying an error in his initial assessment. With one of his original theories, he realized that there was absolutely no mention of either acceleration or gravitation and this was something that he knew he had to address. The issue was further compounded when his close friend, Paul Ehrenfest noticed a peculiarity while watching a disk spinning. With his observations, he noticed that the rim of the disk would spin at a faster speed than the very middle and that, in theory, if a metre stick was placed in the middle then it would shrink.

To the everyday individual this theory would not mean much, but to the likes of Einstein it seemed to complicate matters in his eyes. To him, it meant that certain aspects of geometry must be failing for the disk and this, in turn, led him towards investigating theories surrounding gravity. It was of course an issue he had looked at previously and he would later talk about his anger that he had not been able to initially include gravity in his earliest equations.

The issue of gravity was of course first coined by Newton, but Einstein spent the next decade coming to the conclusion that gravity itself was part of a reality that was substantially bigger than Newton was aware. Indeed, Einstein then wanted to create a theory related to gravity that was connected to the concept of space and time. This would go on to be the theory that would not only completely change the life of Einstein, but change our understanding of the world.

Further Work on his Theory of Relativity.

We mentioned how Einstein had initially produced a paper in 1905 connected to an aspect of the theory of relativity, but this was certainly something that Einstein himself would have regarded as being a work in progress. We are not sure as to how much time he spent on developing the ideas between 1905 and 1912, but what we do know is that he picked up the idea once again in 1912 when he moved back to Zurich for a short time before moving to Berlin.

He was helped at this time by his old friend, Marcel Grossmann although this was partly due to the fact that he was one of the only individuals based in Zurich that Einstein felt could be trusted with his new ideas. This in itself gives us some kind of an insight into the world of physics as there must have been some fear surrounding the creation of new theories and that they would then be stolen as Einstein did indeed keep his cards close to his chest. For Einstein, this fear was something that would actually come to fruition later on in life at the point of revealing his famous theory.

His Move to Berlin.

By 1913, although some say 1914, we see Einstein moving to the University of Berlin in order to take up a position that he was recommended for by Max Planck himself. Indeed, he was not only recommended by Planck as he was actually head-hunted. Furthermore, he had to be granted the position by the Kaizer himself who had to be convinced that Einstein was the correct person for the job, a job that did not even require any teaching if that is what Einstein himself wished to happen. This also coincided with the beginning of his affair with his cousin and it was the start of the failure of his marriage. It is known that at this point his family, as he knew it, would never live together again as his wife had decided to stay at home in Zurich rather than travel with him. This was a situation that was certainly not appealing to Einstein although it was also one that he was hardly desperate to change.

He took up a number of titles during his early years in Berlin with him holding the position of the director of the Kaiser Wilhelm Institute for Physics, the Professor of Physics for the Humboldt University of Berlin, and he was also a member of the Prussian Academy of Sciences. Furthermore, he was also involved in correspondence with Leiden University in the Netherlands although he was primarily focused on a relationship with two of the physicists that were based there by the name of Hendrik Lorentz and Willen de Sitter.

During his time in Berlin, Einstein remained focused on examining the basics of the various principles as put forward by Newton who he held in high regard. He was determined to see how different theories by Newton would fit in with his own theory of relativity, which is of course his most famous work.

However, at this moment in time we will pass on talking about that particular theory until a later chapter where we will examine his various theories and concepts that led us into the modern world of physics. Such is the importance of the theory that it does deserve to be examined in some detail without any interference.

Chapter 6: Continuing in Berlin with More Theories.

Einstein was, of course, subjected to the issues surrounding the outbreak of war in 1914, so his career was interrupted to a certain extent as Germany focused on more pressing matters. He did deal with his Theory of Relativity, but we do know very little about these years due to the fact that a war was raging around him. However, Einstein was a strong pacifist and he did have clear ideas about war and wishing it would end. We will examine this aspect of Einstein in some detail later on due to its importance in his story.

You must remember that Einstein would go on to live in Berlin for close to 20 years. It is interesting to note that he maintained his Swiss citizenship even though he was back in the land of his birth, but he was clearly still concerned about war and being dragged into it even though he was an academic.

What we do know is that early on during his time in Berlin he saw fit to become involved in politics and he is seen as having played an integral role in the development of the German Democratic Party.

It is also believed that his time in Berlin was not the happiest point of his life. By all accounts, there was a sense of frustration coming from him that he had not been able to make any further progress after he revealed his theories on relativity and, at least initially, his relationship with Berlin was purely on a professional basis.

Another point regarding Einstein that does provide us a rare glimpse into other aspects of the man himself was the way in which he appeared to have little sympathy or understanding of those individuals that were close to him. Even after his first wife divorced him, it is known that his second wife still had to contend with him being a womaniser mixed in with periods where he effectively isolated himself in his room. This was his way of working on his theories, but it was certainly not an easy marriage due to the way in which he insisted on living his life.

During his time in Berlin there appeared to be some kind of personality cult that developed around him due to his theories and the impact that they had. He is known to have grown tired of this and wished to have a more relaxed life along with more control. However, he did take up the offer of a house from the mayor of Berlin, which was seen as being a special gift for him turning 50 and bringing so much fame to the university in the city.

At the same time, he also grew tired of spending every single day in the city and that is why he would often spend most of the summer at his very own summer home which was located in Caputh. There, he was known to spend a considerable amount of time playing his beloved violin, sailing, continuing his work on his theories, and rather surprisingly entertaining a wide array of guests that ranged from important political figures to Charlie Chaplin.

However, his time in Berlin is ultimately known for one thing, his Theory of General Relativity.

The Revealing of the Theory of Relativity.

If we can now go back in time slightly to 1915 which would turn out to be the most spectacular year of his life if we are only looking at the different theories and concepts that he created during his career. In the summer of 1915, Einstein was invited to give a number of lectures at the University of Gottingen and it was here where he began to reveal some of his brand new concept related to the Theory of Relativity. For Einstein, this was coming to the culmination of years of work as he had been plagued by the idea of gravity not having been included in his initial theories on the concept of relativity some years earlier.

The thing about Einstein was that he was quite clever in his approach to the revealing of his theories. He understood that it would be debated, doubted and put to the test and he clearly wished to be sure that it would stand up to that kind of scrutiny. For him, there was only one way to go and that was to careful pick and choose what he would discuss until he was absolutely ready. However, even though he took precautions, even Einstein would have been unable to predict how his lecture would evolve.

As a result, these lectures gave an incomplete version of general relativity, but it was still enough to perk the interest of various prominent physicists around the world. However, as we said, Einstein was not yet ready to reveal his masterpiece, so he was careful to leave out some key mathematical aspects to prevent others from not only completing the equations, but then stealing his entire concept which he knew was going to be powerful.

The problem for Einstein was that he did not bank on the actions of the individual that had organized the series of six lectures at the university, the mathematician and his friend, David Hilbert. Ultimately, he would fall out with Hilbert for a period of time purely because of the way in which he believed Hilbert had betrayed him after the lectures.

What we know is that Hilbert, who witnessed the lectures, was able to complete the missing parts that Einstein had left out and he then published a paper on the concept of general relativity in November 1915 a mere five days before Einstein was set to reveal the concept for himself. The issue was further compounded by the fact that Hilbert tried to claim the theory as his own in order to take the credit and this was something that led to a major breakdown in their relationship, although they eventually made up once again.

Regarding this issue, Einstein was quoted as saying:-

"I struggled against a resulting sense of bitterness, and I did so with complete success. I once more think of you in unclouded friendship, and would ask you to try to do likewise toward me."

As you can imagine, Einstein clearly felt as if he had been betrayed by a friend as this was something that he had worked on for a number of years only for Hilbert to try to steal his limelight. Thankfully for Einstein, he was accredited with the work and the development of his theory even though there were attempts for the equation itself to be attributed to both.

The actions of Hilbert are certainly peculiar. He only had the part of the equation that he had worked out along with the information that Einstein had revealed. However, he had none of the backing for the theory and it is difficult to imagine how he could have believed that he would be able to pull off this charade. Indeed, he was also unable to inform anybody of the various influences for the theory, which was certainly not the case for Einstein.

The Influences for his Theory.

It is rather interesting to look at the things that influenced this theory as it is simply not the case that Einstein came up with an entirely new concept in every way. Instead, he took what other physicists had created before him and sought to either prove or disprove various concepts or, alternatively, marry together different ideas depending on the results of his experiments.

We have already stated how the work of Newton was a big influence in Einstein, but what is not as well known is the role that David Hume played in the thinking of Einstein. Hume was a Scottish philosopher who, in 1738, produced a book titled 'A Treatise of Human Nature' and it is known that this was something that did have a major influence on the thinking of Einstein for quite some time.

The thing with Hume was that he held the belief that scientific theory should actually be proven via experimentation and discussing the results. He also held the belief that time was not capable of existing independently from the object itself and this was a view that was also shared by Einstein. Indeed, he is quoted as saying that.

"It is very well possible that without these philosophical studies I would not have arrived at the solution."

What is clear to many is that Einstein drew inspiration from a variety of sources when examining, or even creating, a variety of theories. He was never afraid to look outside of physics itself in order to try to answer some of the most pressing questions regarding the universe.

The Theory Itself.

So, what was this actual theory that has become synonymous with Einstein?

This theory not only challenged a whole host of concepts, but it completely altered the way in which physicists largely viewed the world. However, in order to understand the theory in its most simplistic form, we need to also understand the concept of gravity.

It was, of course, Sir Isaac Newton who first came up with the theory of gravity and that two objects produce a force of attraction that pulls them towards one another. However, the pull depends on the size of the mass or object as well as the distance between the two.

If we use the earth and humans as an example, then we all feel the tug of gravity on our body as it is the thing that keeps our feet on the ground, but at the same time we are also pulling away from the body of the earth. The problem for us, although it is clearly a good thing, is that the force that we place on the earth is smaller in size than the pull that the earth has on us. In other words, it is similar to a strength competition and the earth will win with us every single time.

Einstein had a few issues with a number of theories that were put forward by Newton and we have to, once again, state how he had been working on this theory for a decade before it was fully revealed to the world. He had already determined in his theory on Special Relativity that the key laws of physics are the exact same for every non-accelerating object and that the speed of light also remains as a constant no matter the speed that the observer of the light is travelling at.

Additionally, he also looked at the idea that both space and time are actually interwoven into the one space-time continuum and this was seen as breaking the mould as far as physics was concerned. In short, he theorized that events that occur at the same time could, theoretically, occur at different times for other people.

This theory had turned things on their head, but it is clear that Einstein was not exactly finished at changing the way in which these theories were being viewed. Indeed, it was only the beginning when you then consider the rest of the theory that would then evolve.

Changing to the Theory of General Relativity.

Prior to actually discussing this theory we should remind the reader that it is actually two theories combined. The paper published in 1916 took ideas that Einstein had developed in 1905 and married them with further developments he had been working on for a number of years.

While working on his Theory of General Relativity, he came to the conclusion that vast objects were able to create some kind of distortion in what he referred to as space-time. As an example, it is best to imagine what would happen to a trampoline if a large and heavy object was placed in the middle. It would automatically pull the trampoline down in the middle and then if you placed a ball on the trampoline and spun it around the edges, it would be automatically pulled to the middle and towards the heavy object.

This can then be likened to the way in which a planet pulls on asteroids in space with the planet being the object at the middle of the trampoline and the asteroid being the ball that is sent spinning around the outside.

After coming to the conclusion in 1905 that the speed of light is unchanging, Einstein was able to come to the realization that there is absolutely no frames of reference when it comes to the universe. Everything within it is moving relative to every other thing. To Einstein, this in itself was an important leap forward due to the way in which it began to allow other ideas to formulate in his mind.

He took his theories further by coming to the conclusion that both gravity and motion are capable of affecting both time and space whereby the pull of gravity in one direction is the equivalent of the acceleration of the speed of the object in the other direction. At the same time, he also summed up that if light is capable of being bent by the process of acceleration, then it stands true that it could also be capable of being bent by gravity.

The way in which time and acceleration is viewed by others is also something that is unique to the individual according to Einstein. He came to the conclusion that time does not pass at the exact same rate for every individual. For example, somebody who is classed as being a fast moving observer will be able to measure the passing of time at a much slower rate than an individual that was stationary.

This was regarded as being a real leap forward in the understanding of the correlation between time and space and it introduced an idea called time dilation. However, the theory was not purely limited to dealing with time and space. Instead, he also came to the conclusion that an object that is moving faster will always appear to be shorter in size as it passes along than the same object that is moving at a slower speed. The problem is that this difference is rather difficult to see until the object is moving close to the actual speed of light at which point it becomes far more obvious.

Incorporating $E=mc^2$ Into the Theory.

We do of course know that Einstein was responsible for the above equation which he created back in 1905 although he still sought to see how it could be incorporated into this particular theory. In other words, he saw this equation as being representative of a link between mass and energy, but he was also unsure as to how exactly it could be used in the grander scale of things.

Einstein showed that both mass and energy are actually belonging to the same thing although they are different aspects. In the decade that it took Einstein to work on the theory, he then came to the conclusion that an object that is moving at a fast speed will appear to have increased mass in relation to the same object that is at a slower pace. The reason why this is the case is because of the simple fact that by successfully increasing the speed of an object you are also increasing its kinetic energy. By doing this, you therefore increase its mass. Sound complicated? Well, for him it was a logical point to make although it was also one that was extremely important.

What Einstein then did was to state that this was the reason why it was impossible for matter to travel faster than the speed of light. Why is this not possible? Simply because of the fact that the mass has to increase with the speed and if this continues to happen, then the mass itself becomes infinite when it reaches the speed of light. At this point, it requires infinite energy and this is the part that just cannot happen.

Going Back to his Space-Time Continuum and Gravity.

As you can see, his theory surrounding general relativity is rather complex as it does include a variety of his theories and combines them into the one thing.

However, from the point of view of gravity, there was one major flaw in the theory as put forward by Newton in that he was unable to explain where it came from just that it actually did exist.

That changed thanks to Einstein as he was able to show that relativity was capable of explaining where gravity originates and that origin is the way in which objects that have a large mass are capable of warping that space-time continuum. This was something that was given the name of a gravity well and the best way in which it can be explained is by looking at objects that are orbiting another mass.

Generally speaking, the object should follow the shortest path possible as that will also lead to it using up the least amount of energy. For example, if you think about a planet, then it generally goes in an eclipse around the sun. This is the shortest route, but at the same time it is also working within the afore-mentioned gravity well of the sun.

Sounding Complicated.

This does sound rather complex, but the theory itself can be explained in simpler terms particularly for those individuals without a background in physics.

If you can understand that the speed of light remains constant for everyone along with the concept that observers moving at a constant speed will also be subjected to the same physical laws. From this, Einstein created his theory that time must be capable of changing according to the speed of an object that is moving in direct reference to how an observer is seeing it. This formed the basis of his 'Special Theory of Relativity'.

This, to many, sounds impossible, but it is something that has been shown to be true thanks to a number of experiments carried out by scientists since Einstein created the theory. One such example is an experiment whereby it was shown an atomic clock does indeed tick more slowly when it is moving at high speed. Of course, at the time of his theory everything was just that, a theory, and it took a number of years before various aspects were scientifically proven to be correct.

In a nutshell, Einstein created a theory that both time and space are actually relative to one another (hence the name of the theory) when dealing with what is seen as being a special case, the absence of a gravitational field.

This part alone was something that shook the world of physics to its core, but as we have previously stated Einstein wanted to take it further and to incorporate gravity into his theory and that is the part that took 11 years for him to resolve. In his opinion, everything in the universe should be able to form a part of the same equation or theory, which would ultimately lead to him working on a unified theory without any success, and the next step was to include gravity.

Eventually, he concluded that matter is able to cause space to curve, which we mentioned a short while ago and that, therefore, gravity is a curved field. You may recall that this went against the ideas of Newton, but Einstein still concluded that it referred to a part of space that was under the influence of this force that is created by the presence of a mass.

Exactly Why is it Important?

To the everyday person there would probably be a question as to why such a question is important, but the answer is simple. This theory changed how the universe was viewed by scientists and it immediately altered our understanding of it. We will even discuss how this theory has had an impact on our daily lives later on in the book as it did so in ways that most people would never link with Einstein.

Up until this point everybody had followed the concept of Newton and his ideas of gravity without any hesitation although to Einstein it was clear that Newton and his theories did not allow for the way in which planets would orbit around the sun. This is now seen as a basic theory surrounding our universe and Einstein held the belief that our universe was actually a fixed and static entity although his later work appears to contradict certain aspects of his very own theory although this is hardly unique in the world of physics as theories evolve.

Indeed, it eventually reached the point that Einstein came to the conclusion that the universe was in a permanent changing state, something which was supported by an American astronomer by the name of Edwin Hubble who would, of course, go on to give his name to the famous telescope that peers into the darkest recesses of the universe. As we will mention later, this conclusion links Einstein to one of the other most famous aspects of physics, the 'Big Bang Theory'.

It should be noted that for our everyday life that the idea put forward by Newton was certainly more than adequate for explaining how things affect us here on earth. However, the theories by Einstein took things to a new level and in explaining the bigger picture of what goes on beyond the realms of planet earth which, as a result, does have an impact on our daily life even though we are not initially aware of the impact.

As an interesting side note, the theory as put forward by Einstein has since been used to great effect by the science fiction community who have used ideas of black holes, parallel universes, and wormholes for their stories. Indeed, even Star Trek would be vastly different, or not even have existed in the first place, if it had not been for the work of Einstein as the concept of warp speed would have probably not even been imagined by creative writers. Once again, we will look at the impact that was had by Einstein when we examine other aspects of his life and popularity later on.

How He Proved It.

To Einstein the best way to show an example of his theory was to refer people to the way in which you could measure the angle of deflection of light from a star that is traveling close to the sun. Thanks to the strength of gravity, the angle of deflection is twice what is expected. Furthermore, to Einstein it also explained why the light from stars that are affected by a stronger gravitational field will emit light that is closer to the red end of the color spectrum than those that are not being influenced by the same force.

The difficulty for Einstein was that he was a theoretical physicist. In other words, he would put up a theory and explain it according to equations as to how something should work in theory, but it was then up to others to go ahead and prove it. This was something that had happened to him throughout his career and that is why it then took a number of years for things to be proven as it was left in the hand of other scientists to not only understand the theory, but then understand how to show that he was either correct or wrong in his thinking.

After the Theory.

One of the things about Einstein was that he never really rested on a theory once he had published findings. This can be seen by the way in which he spent 11 years trying to incorporate gravity into his initial theory, but even after achieving this he would go on to spend 30 years trying to take it even further.

For him, the next question related to this theory was to resolve the question of a unified field theory whereby he would then be able to include every property of both energy and matter within a single equation. This idea would plague his thoughts as he searched continually for an answer to something that, for him, would hold the key to so many aspects of the universe and how the world in general operated.

However, Einstein did come across a problem in that this entire concept was being complicated by the what is known as the 'Uncertainty Principle' of quantum physics as this would regularly stop him in his tracks. This principle stated that it was impossible for the movement of just a single particle to be measured accurately for the simple reason that neither the speed nor the position could be measured with an adequate degree of certainty. This would ultimately become one theory that was beyond even Einstein although he did manage to push forward the debate to such an extent that his work has certainly made it easier for physicists that have followed to continually search for these answers even though, at this point, it remains unsolved.

This work can be linked to the concept of a brand new science that became known as cosmology, which has seen a number of developments over recent decades. It would not be possible had it not been for the work of Einstein.

Chapter 7: The Theory Brings Him Fame.

Even though he had been making waves in the world of physics for a number of years, it appears that the 'Theory of General Relativity' really propelled Einstein to a whole new level. Previously, he had been invited to speak at a number of different universities across Europe, but it seems that after the end of World War I that he then began to travel further afield as scientists around the world became intrigued by his various concepts.

It is important to remember that even though he created his famous theory in 1916 that it took another three years before it was successfully shown in experiments. It was at this point that his life changed and it certainly set the scene for him to become known as perhaps the most famous physicist in the world. The theory itself was eventually proven be correct by British physicists who traveled to a remote island in Africa to view the eclipse of the sun. In doing so, they were able to calculate that Einstein was indeed correct with his theories regarding light with it being at this point that his concept was actually accepted by the scientific community, although not entirely.

The reason why this was such a big deal is clear for all to see. It was the first major breakthrough in understanding gravity for approximately 250 years and it managed to completely change the way in which the universe was viewed. In the process of tackling his theories, he had managed to alter the way that people viewed Newton and his work, although he was still credited by Einstein as being a major influence. Considering the high regard that Newton was held in, it was certainly something rather special for him to be usurped by an individual who was still relatively unknown apart from within the realm of physics.

You only have to look at the media from the time to see that there was hardly a magazine in existence that did not mention him. Indeed, you only have to look at the way in which The Times of London proclaimed his theories as being a 'revolution in science' to see how even normal newspapers were enthralled by the ideas. To Einstein, the entire idea of fame surrounding this theory led to him stating that he preferred to refer to it as being the 'Circus of Relativity' which does show his real displeasure at how things had developed.

This was a scientist who had previously experienced some fame due to his earlier theories, but it was primarily connected to being invited to talk at various universities around the world. This time it was different. This time he would experience fame amongst normal people who suddenly had a real interest in the world of science like never before.

1921 – Nobel Prize Winner.

As a result of his achievements related to his theories, Einstein was awarded the Nobel Prize for Physics in 1921 although he was unable to collect his award until the following year. However, it is interesting to note that as there were still various scientists that had issues with his theory that the prize was not actually related to the specifics of the Theory of General Relativity.

Instead, he was given the award in recognition of his work on the photoelectric effect although Einstein still decided to discuss the concept of relativity when giving his acceptance speech. This could, in part, be attributed to the fact that Einstein was actually a bit of a rebel and was making a point that he was being rewarded for one thing and yet his most important breakthrough was being ignored at this point.

The reason why the most famous theory of the history of physics was ignored for the Nobel Prize? Simply because it was still regarded by a number of people as being rather controversial, but the committee felt that they could not ignore Einstein partly due to his popularity. In this sense, the prize can perhaps be viewed in a slightly different light as a way to pacify people who would have been confused as to why he was not given this award.

1921 was also the year when Einstein traveled to the United States for the very first time. This was a major sign that he was being courted by countries around the world, but it is significant for the story of Einstein as the United States would go on to play an important role in the rest of his life, although not for a number of years yet. Indeed, he did appear to initially fall in love with the country upon his arrival.

The Years After the Prize.

Throughout the course of the rest of the 1920's, Einstein certainly suffered from a sense of being unable to match his earlier successes. That is not to say that he was guilty of doing nothing apart from touring the world as that was not the case. Instead, he was beginning to become too consumed with other theories that would ultimately prove to be too difficult even for the great man himself to crack.

That does sound as if he contributed nothing for a number of years, but that is also not true. He did continue to make smaller advances in the world of physics and in particular he worked closely with an

Indian physicist by the name of Satyendra Nath Bose and between the two of them they created the concept of a brand new type of matter, which became known in scientific circles as the Bose-Einstein condensate. Furthermore, this theory dealt with a type of matter whereby a superfluid can act like a liquid, but it has no viscosity. Previously, it was assumed that this meant it was impossible for it to be a liquid, but the two physicists managed to show that this was not the case.

We do know that Einstein would spend these years becoming embroiled in mathematical issues rather than making any substantial progress. He also took the stance of largely ignoring quantum physics, which was the major movement of the 1920's. This was of no real interest and he also doubted various aspects and he became rather disillusioned of physics in general. As a result, his isolation increased putting a strain on his work elsewhere.

By December of 1932, Einstein had decided that enough was enough and he made the decision to emigrate to the United States and to leave his position at the University of Berlin. Instead, he would take up a brand new position at Princeton, New Jersey where he would focus on the Institute for Advanced Study, which was an entirely new department that had just been founded. However, we should mention that his hand was effectively forced by the rise of the Nazi Party, a point that will be mentioned again later.

His Move to The United States.

Einstein moved to the United States at the beginning of 1933 and his time at Princeton allowed him to focus his attention on the idea of the unified field theory. This was something he had begun to investigate some years earlier, but it was now that his attempts to produce a single framework that could incorporate the various laws and rules of the universe really took precedence.

The United States was a country that had held some appeal for Einstein for a number of years as he had been enthralled by it on his first visit in 1921. It should be mentioned that the US press were also excited about Einstein when he appeared on its shores for the first time as he was described as having a rather distinctive appearance and had the kind of character that they loved. He was certainly not your usual kind of scientist or academic. Instead, he was regarded as being rather humble and warm to people whereas they were previously used to scientists being rather aloof and believing that they were simply more intelligent than others and, therefore, should be given the respect that they deserved.

Einstein was also loved in the United States due to his ability to provide quotes that would then be picked up by the media. They simply loved his small quips that were so memorable and they certainly went against the grain of scientists who were regarded as being too stuffy as individuals to come up with anything quite so memorable. He even became famous for never wearing socks and answering his door in bare feet even when he had important guests coming. Considering he was not exactly mad, he made a good job of trying to convince people otherwise.

Indeed, Einstein clearly felt the same way in return as he was quoted as saying how friendly and open the country was and that there was also a real sense of optimism about the country in general. He settled there quite quickly with him taking up his post at Princeton although it was not always plain sailing for him.

Einstein was faced with the death of his second wife in 1936 and this is known to have devastated him. You must remember how he had been with her for some 20 years after they had begun their relationship by having an affair and she had been with him through some tough years and also an increase in his level of fame. You should also remember that she stayed with him even though he had a roving eye. However, he now found himself largely being alone and what made this situation even worse was the fact that he had already started to isolate himself from the scientific community as well due to his pursuit of a unified theory. His colleagues would, instead, focus on other areas of physics including quantum theories with this bringing Einstein into conflict with them and increasing his sense of isolation.

For a number of reasons, the 1930's were not the best decade for Einstein and even though we will examine how he focused on world peace through his pacifist links, we should also mention one other key concern that would plague him for the rest of his life.

Einstein and the Atomic Bomb.

By the time we move closer to 1939, Einstein had been shown to have been quite correct in moving from Nazi Germany as World War II was on the brink of becoming a reality. However, even though this was clearly of concern to him as a pacifist, he had a more pressing issue on his mind.

It is known that his famous equation $E=mc^2$ was eventually used to help further the advancement of the concept of the atomic bomb and there is no doubt that Einstein was aware of this possibility from an early stage. Ultimately, it led to him contacting President Roosevelt in 1939, after being pressured by Leo Szilard, the Hungarian physicist, to encourage him to develop the idea of the atomic bomb although not for actual use. To Einstein, it was important that the United States was able to show that it had the upper hand as an attempt to diffuse the potential deadly situation that would arise should Hitler manage to get the atomic bomb first.

The one area where Einstein made a mistake was that he believed the United States would never then go ahead and use the bomb if it had the technology itself. Also, he would not help them to build the bomb as that was not what atomic energy had been created for in his mind and as a staunch pacifist it would destroy his own ideas regarding peace.

Indeed, a number of scientists had left Germany as they feared being dragged into its development and they wanted to have nothing to do with it, but there were still some physicists back in his homeland that certainly had the capabilities. Einstein strongly believed that if a country was able to show that it had the bomb, then the others would become too afraid of what could happen that it may even lead to the end of the war as a result. As it turned out, this was perhaps one of the only ideas that Einstein had that was perhaps not that sensible although this is only through hindsight.

You must remember that two German scientists, called Otto Hahn and Fritz Strassman, had already shown that the creation of an atomic bomb was at least a theoretical possibility in 1938 as they had worked on the area of nuclear fission. Their ideas were then further developed by another two scientists called Otto Frisch and Lise Meitner, so things were certainly gathering pace. You can even understand why Einstein was so concerned about this as he could easily see where all of this was heading.

His arguments clearly held some water in the United States due to the level of respect that he had and it did ultimately lead to the creation of what became known as the Manhattan Project. This was the research and development project that was led by the United States along with contributions from both Canada and the United Kingdom that did ultimately lead to the creation of a nuclear weapon.

However, Einstein himself refused to take part in the project, although clearly he would have been asked and viewed as being a major asset, as he was a strong pacifist and felt that this went against so many of his ideas. For him, it was enough that he had managed to convince the United States to begin to develop the concept of the atomic bomb even though he hoped that it would never be used in anger. Indeed, as a pacifist he was strongly against it being used as a weapon and even signed a manifesto declaring the dangers of nuclear fission as a weapon along with the British philosopher Bertrand Russell.

The surprising thing was that his plans and desire for the United States to follow this path did not go down well with everybody within the country. It is known that J. Edgar Hoover of the FBI had a particular interest in Einstein and he was placed under scrutiny as Hoover felt that he could not be entirely trusted and believed that there must have been some other ulterior motive behind his desire for the US to go down this path. However, those suspicions were unfounded.

How the Letter Developed.

The way in which this letter developed provides us with some insight into the way in which Einstein was viewed by the scientific community. It is clear that he was viewed with a certain degree of respect from his peers and his experience with other prominent individuals was something that other scientists wished to take advantage of.

It was known that Hitler had been looking towards the prospect of building a nuclear bomb and indeed there had been several meetings of prominent scientists in Germany to discuss such a matter. These discussions were supposed to have been behind closed doors, but news of them leaked out.

Scientists outside of Germany, and in particular within the United States, became alarmed at the amount of information that the Nazi Party had allowed to come out. Their reasoning was that they must be well on the road to building such a device if they allowed the media to report it.

The fears of the international community were further enhanced by the way in which the Nazi Party stopped the sale of uranium from Czechoslovakia. This could only mean one thing, but there was one important point that had been overlooked by the Nazi Party and that point was that the best uranium for this process came from Belgian Congo. The fact that the Nazi Party were looking at invading Belgium was certainly something that alarmed scientists who then believed it would only be a matter of time before they obtained the bomb.

We previously mentioned Einstein being contacted by a Hungarian physicist called Leo Szilard who had previously been a student under Einstein several years earlier. However, his reasoning for contacting Einstein was due to him being aware that Einstein was a personal friend of the Belgian Queen Mother. Szilard hoped that he would be able to use this influence to encourage Einstein to write to the Belgian royal family to convince them to prevent the uranium from falling into the hands of the Nazis. Einstein was only too happy to help.

By the time we get to 1939, Szilard had decided to visit Einstein at Princeton and this time he brought with him another two scientists, Eugene Wigner and Edward Teller. They were similar to Szilard in that they were both Hungarian physicists who had fled their homeland and were now effectively refugees. By this point, Szilard had become even more concerned with the potential for a nuclear bomb that he had another idea that he wished to put to Einstein.

With him on his travels was a draft of a letter that he had hoped would land on the desk of President Roosevelt. However, he also knew that nobody would pay attention to a relatively unknown Hungarian physicist, but he was also aware that people would be more likely to listen if Albert Einstein was involved. Upon reading the draft, Einstein made a few minor changes before willingly putting his name to it.

The contents of the letter are well known and the main idea was that it highlighted Einstein's fears as to what was going on in Germany and where it would lead. The letter eventually ended up in the hands of the economic advisor to the President, Alexander Sachs who hand delivered it to Roosevelt immediately. His reaction upon reading it? Absolutely no hesitation in the need to act on what was said in the letter leading to the Manhattan Project that we mentioned earlier.

Finally, it is worth noting that after the US dropped the atomic bombs in Japan that Einstein commented on how signing that letter was the biggest mistake that he had ever made in his life and one that he regretted for the rest of his days. The fact that his original E-mc2 equation was also used in the creation of the bomb, albeit at the earliest stage, was also something that rankled with him and it played a role in him pushing for scientists to take more responsibility for what their discoveries could eventually be used for.

By 1940, Albert Einstein had decided to become a US citizen due partly because of the way in which he had been accepted by the country and the simple fact that he was unable to return to his homeland. It is unknown if he then regretted that decision just a few years later after the bombs were dropped.

Regarding his time in the United States, he was certainly impressed with some aspects of the country and its society, but rather disappointed in other areas. He was quoted as saying how he was impressed with the way in which it led the way with technology and he understood the amount of influence the country had on international politics.

However, he was disappointed in the way that the average American simply did not take any notice of what was going on around the world. He saw them as remaining idle and just not willing to take part in making changes that he felt the United States was more than capable of doing around the world.

Chapter 8: His Theories Summarized.

Considering the importance of his various theories, and the fact that we have mentioned them in a number of places throughout this e-book, it is a good idea to summarize them once again in their own chapter. We shall also, where we can, look at the way in which Einstein managed to create the theories in the first place as this gives us some kind of an insight into the man himself.

However, we should note that it is impossible for us to go through every single theory or scientific paper that he was responsible for during his life. It is estimated that he produced over 300 such papers and even another 150 that were non-scientific in nature, so this was a man that was extremely busy in a number of areas.

At this moment, it is perhaps best that we focus on his primary theories as well as any additional comments that he made on the work of others.

The Investigation of the State of Aether in Magnetic Fields.

In an earlier chapter, we mentioned this paper as being one written by Einstein when he was only 16 years old. We know that upon completing it that he then sent it to his uncle, called Caesar Koch in order to get his opinion on his ideas. Sadly, we are not aware of the actual extent of the paper or even what the reaction of his uncle was to it, but it is perhaps enough to say that clearly it cannot have been too bad as it certainly did not stop him from continuing along this particular road as an adult.

Even though we know little about the paper it does show that Einstein had already started down this particular road and it gives us a glimpse into the actions of the genius at work at an early stage.

On A Heuristic Viewpoint Concerning the Production and Transformation of Light.

This was the first paper that Einstein published in 1905. It focused on the photoelectric effect and considering it was one of his earliest papers that was published in any seriousness, it had a profound impact on physics in general. As we mentioned in an earlier chapter, this paper dealt with theories put forward by Planck with the two going on to become firm friends along with the sharing of ideas.

This paper focused on several key points although it did spend time looking at a particular puzzle regarding energy. Einstein was able to prove that it is only ever exchanged in discrete amounts, known as quanta, and as a result he was able to help lay the foundations for what became known as quantum physics. It is then quite ironic that while he later spent time on his unified theory that he largely gave up his pursuit of quantum physics with him then getting little support from the rest of physics due to them focusing their attention elsewhere.

We can look upon this paper as being a wonderful example of how the work of Einstein was largely viewed with a sense of suspicion for quite some time. In the paper, he created the theory that light was indeed made up of particles called quanta. This one single point was then largely ridiculed by physicists around the world who poured scorn on the idea that such a thing existed. It took 14 years before Einstein was proven to be correct in his theory.

On the Motion of Small Particles Suspended in a Stationary Liquid as Required by the Molecular Kinetic Theory of Heat.

This was the second paper published by Einstein in 1905 and as was stated in an earlier chapter, this focused on what was known as the Brownian Motion. However, this paper is important as it added to the theory of the existence of statistical physics as well as adding to what was already known about the atomic theory. The information that was uncovered by Einstein in this paper would ultimately play a part in the development of atomic energy, which would then be used in a way that directly went against the wishes of Einstein. However, in support of him there was no way that he was able to predict that the use of atomic energy would evolve in the way it did.

On the Electrodynamics of Moving Bodies.

This was the third paper and it was published two months after his second paper. This time, he focused on his concept of special relativity which would go on to help form the basis of his theory of general relativity later on in his career.

We previously noted how this paper focused on the work of Maxwell and in particular his equations related to both electricity and magnetism. This paper was viewed as being important as it sought to answer questions related to the speed of light and issues related to laws of mechanics. He took the concepts that had been put forward by Maxwell and expanded on them even further while the paper also sought to discredit the concept of the luminiferous ether.

Does the Inertia of a Body Depend on its Energy Content?

The fourth and final paper of the year may have a rather uninspiring title, but it did lead to E=mc2. The paper itself was published in November of 1905 and it also helped to lay the foundations for nuclear energy, which is another point that would largely come back to haunt Einstein towards the end of World War II. In this paper, he sought to prove the existence of what he referred to as being 'rest energy'. However, the main focus was on the area of matter and energy hence the famous equation being created. Furthermore, the paper sought to prove that gravity could indeed bend light with gravity itself playing a key role in further studies by Einstein.

Considering the way in which this one single theory has become so famous and memorable, this paper did not make major inroads within the world of physics.

Contributions to Other Theories.

Einstein was also known to have contributed to other theories both directly as well as indirectly. Often, his own observations would lead to breakthroughs for other individuals who were already studying certain areas and yet Einstein with just a comment could help them out enormously.

For example, his observation in 1911 of the dual nature of the world due to the co-existence of both waves and particles within quanta really spawned the beginning of the development of the entire study of quantum physics. It had previously been put across as a theory, but after this point more physicists would find themselves being drawn into this area even though Einstein himself did not feel that same urge.

Furthermore, Einstein returned to concepts of the quanta in 1919, which was a full six years before the creation of the idea of quantum mechanics, when he came to the conclusion that there may very well be an issue with the dualism of cause and effect. To him, he believed that it may be impossible to link effects to causes within quantum physics. As a result, his words had their own dual effect with some physicists being put off trying to link them while for others it set a challenge that they wished to attempt to meet.

Staying with the quantum theory, Einstein visited it once more between 1924 and 1925 although this would largely be the last time that he would be drawn into this part of physics. Considering the actual implications associated with this part of physics, it is a surprise that he believed it to be nothing more than a nuisance to him. However, it is interesting to note that when the quantum theory was effectively in limbo between 1900 and 1905 that it was only Einstein that effectively took up the mantle in order to push the theory forward in accordance with the work conducted by Planck. The problem was that even Planck was skeptical of his own theories surrounding quantum physics although Einstein showed more confidence in it with him using aspects of hypothesis put forward by Planck in his own theories that formed his work in 1905.

What we know is that comments made by Einstein on the quantum theory in 1925 spawned a resurgence in quantum mechanics as a field of study with considerable attempts being made at advancing understanding and knowledge between 1925 and 1927. However, this was the point where Einstein effectively stood back from it all in order to concentrate on his own work, which were mainly focused around the concept of the unified theory.

What is being said here is that Einstein appears to have started off his career by creating a number of theories, but his career then began to slide in the 1920's although that is largely down to his own issues. He may have produced a number of scientific papers, but they mainly focused on smaller aspects of physics and actually bare no real impact on our story.

However, there is one area that we still need to examine and that is what became known as the Unified Field Theory.

Einstein and the Unified Field Theory.

Einstein was always the type of individual who could never just rest on his previous achievements. To many, coming up with the Theory of General Relativity should have been their crowning glory, but Einstein was still not entirely happy with his work. Instead, he sought to create a unified field theory which would incorporate both gravity as well as electromagnetism into the one equation. This was something that he would continually work on over the course of the last thirty years of his life.

To him, there was an inherent need to try to unify the laws of nature into the one equation although he clearly understood that this would not be an easy task. He held a strong belief that there was a link between aspects of quantum mechanics along with electromagnetism and gravity. He held the belief that it was just implausible for there to not be a theory whereby everything was combined as one. This belief was further enhanced by the way in which he never felt satisfied about the apparent randomness of the quantum theory and the entire concept troubled him quite considerably.

When Einstein began work on his unified field theory, physicists only knew about the existence of protons and electrons as well as gravity and electromagnetism. The idea of quantum theory was starting to make some inroads and a number of physicists that were involved in this field were rather excited at what it could hold. To them, it was pointless trying to combine it with the other forces that were already understood as this was something new and, to an extent, focusing on anything else was a waste of time.

However, Einstein was not the only individual that had decided to start work on a unified theory. Indeed, one of the first to work in this area was a scientist called Hermann Weyl who began to formulate theories in 1918. The ideas that were put forward by Weyl were then expanded upon by another scientists called Theodor Kaluza who looked at the space-time concept as put forward by Einstein and developed it further. He spoke about the space-time dimension as having five distinct dimensions, which was one more than Einstein, with the extra set being linked to ideas to do with electromagnetism as proposed by Maxwell.

This approach by Kaluza was something that appealed to Einstein and he was intrigued by the idea of five dimensions. He is known to have written to Kaluza effectively congratulating him on the theory and stating how he would have never have come up with this concept. Kaluza then published his paper on this aspect of the unified theory in 1921 with Einstein then including a follow up in 1922.

One thing we know about Einstein is that he was quite open to exploring more than one theory at any given time in order to reach his end goal. He may have liked the approach by Kaluza, but he also felt it worthy to explore other avenues in his attempt to create his united theory. This time, he went back to his own idea of general relativity and tried to expand upon it so that the various equations related to electromagnetism could perhaps be incorporated into the theory. However, even though he worked tirelessly on both potential methods, he could never come to a satisfactory conclusion even after thirty years.

There is a real sense of disappointment in Einstein that he failed to achieve this particular aim. Indeed, he is quoted in 1938 as stating that most of his ideas ended up being rejected very early on before they were even allowed to develop to any real extent as he simply rejected things on a continual basis.

What was Behind the Failure?

Since the death of Einstein, a number of scientists have tried to get to the root of why he failed to achieve this unified theory. For some it is easy in that they believe that it could be impossible for the human mind to resolve it, but for others the key was perhaps in something that Einstein appeared to reject.

That point is focused on the way in which he discredited quantum mechanics. As a result, he is known to have moved away from the main physics community and because of the distance he put between himself and others it did lead to him becoming alienated and not keeping up to date with the latest advances. It has also been pointed out that he would become embroiled in mathematical issues and arguments later on in life rather than the more free-spirited approach that he preferred when younger. This, to some, resulted in him becoming more closed off to various theories that may very well have led to a breakthrough.

Alternatively, those that are strong supporters of Einstein simply state that the reason why he failed was because other tools that he would have required to create this unified theory just simply did not exist before his death. He was ahead of his time to such an extent that it was his superior knowledge and the lack of others being able to keep up with him that caused the problem. At least in some way this does make sense although to which extent is up for debate.

To this day, the unified theory has still not been proven although a number of physicists are still trying to achieve this holy grail. Indeed, we can look at what is known as the string theory as being a possible solution although it is seen as still being something that is far beyond our current understanding.

Other theories connected to Einstein were certainly rather less prominent in the second half of his life. We can presume that this was a direct result of him focusing on the Unified Theory although he did pass judgement and comment on work of other individuals that would have an influence on what he himself was doing.

Chapter 9: Retirement from Academic Life.

Einstein gradually moved away from academic life as we progress through the war years although he did not let up in his pursuit of proving more theories or even creating entirely new ones. Indeed, it could be argued that he never actually retired since his thoughts and theories were still sought after by universities and other physicists around the world.

He left his post at Princeton University in 1945 and for the rest of his life he preferred to indulge in various passions including sailing as well as continuing his pursuit for peace and a new level of international cooperation. He did also largely become a recluse, at least compared to previous years although he was still being asked to give lectures and you cannot forget the fact that he was also asked to become the President of Israel, although this was merely a symbolic gesture.

Less Scientific Work but a Focus on Peace.

After leaving his post at Princeton University, Einstein did cut back on his scientific work, but he appears to have not let up on his ideas of advancing the peace process. One of his primary focus areas was on the idea of there being an international government that would have its roots firmly entrenched in international law. To him, there had to be this sense of international cooperation, which was hardly a new concept although previous versions simply lacked the authority.

Furthermore, Einstein worked tirelessly in opposition to the United States attempting to create the hydrogen bomb, which would have been even more powerful than the atomic bombs that were dropped on Japan. Einstein had been appalled at the United States using this weapon in the first place as he had mistakenly believed that they would have been respectful enough to avoid doing this, hence his letter to Roosevelt. The very idea of a bomb that was even more powerful being created by the very nation that had dropped the atomic bomb was certainly something that filled Einstein with dread.

Considering the size of the issue that he was faced with, Einstein certainly had no interest in backing down. Instead, he believed that he did indeed have the correct solution although it was something that nobody would have ever expected him to do.

The Cold War and Dealing with the Soviet Union.

In what is a rather intriguing twist, and a sign of how much confidence he had in his own ability, it does appear that Einstein made attempts to prevent the escalation of the Cold War. His fears were firmly rooted in the race for more nuclear weapons, something that he was firmly against, so it is perhaps less of a surprise that he felt he could approach the Soviet Union in order to try to come to some kind of reasoning with them. When you think about how world peace was on a knife edge, it is remarkable to think that a humble scientist, albeit the most famous scientist in the world, believed that he was in a position to make a difference.

To even stand the smallest of chances of being successful, he had to attempt to build bridges with the Soviet Union. To achieve this, he made it publicly known that he was against the persecution of Communists by the United States. He also directly opposed the idea that was put forward by both the US and UK for the rearmament of Germany so soon after the war. In this instance, it was not the rearmament himself that he was arguing against, but more the fact it was a direct response by the west to attempt to counteract the rise of the Soviet Union.

Even though we know little about any dealings directly with the Soviet Union, it does appear that this stance had at least some impact back in the United States. Indeed, he was referred to on a number of occasions as being a Communist sympathizer and it did change the opinion of various individuals. This was something that did upset him as he felt that public opinion of him had changed slightly and he had previously viewed the US as being extremely friendly towards him.

What we do know is that Einstein did fail in an attempt to stop the Soviet Union from also becoming a nuclear nation when they exploded their own atomic bomb in 1949. The following year, Einstein appeared on a television show that was discussing the prospect of a hydrogen bomb being developed. With this, Einstein had some very clear ideas.

"The belief that it's possible to achieve security through armaments on a national scale is a disastrous illusion. The arms race between the US and Soviet Union assumes hysterical proportions. On both sides, means of mass destruction are being perfected with feverish haste and behind walls of secrecy. Radioactive poisoning of the atmosphere is now possible. But our goal should in fact be to do away with mutual fear and distrust."

What we see here is that Einstein had a better grip on the actual root causes of the Cold War than perhaps even the politicians of the time were able to achieve. He realized that it was linked to a fear of one another, so to him this was something that could actually be resolved. However, he was still unable to fully understand what was going through the mind of both countries.

Einstein and Israel.

Due to his Jewish roots, Einstein had a close interest in the development of Israel after the end of World War II. He had previously spoken out about the concept of Israel before it was then created in 1948 although he was quick to point out that the only way in which it could exist was if both Jews and Arabs were able to live peacefully together. This in itself was seen as being an admirable opinion to have and indeed it was an opinion that he held even after the creation of Israel.

Of course, the fact that he was one of the most prominent, and popular, Jews in the world did push him into a stronger position after it was created and we know that he was invited to work at a university in Israel even after his retiral from the position at

Princeton University. However, in 1952 he was invited to become the second president of Israel, a move that certainly took him by surprise even though he then had to reject the offer.

"I am deeply moved by the offer, and both saddened and ashamed that I cannot accept. But I lack both the natural aptitude and the experience to deal properly with people and to exercise official functions. I am the more distressed, because my relationship to the Jewish people has become my strongest bond."

It does appear to be the case that Einstein did not even contemplate taking up the offer at any point. He was simply too aware of his own failings and realized that he was not suitable for the job even though it was more a ceremonial position rather than anything else. It is difficult to see Einstein in this position and not using it to then further various causes and he would have understood this about himself leading to him becoming rather reluctant to take up the role of President.

His life after retiring from academic life was certainly busy, but most of what he became involved in was connected more to his work as a pacifist. This part of his life was certainly more important than most people are aware of and that is why we will study it in more detail later.

Chapter 10: Death and Studies of his Brain.

By the time we move into the 1950's Einstein himself knew that his health was beginning to fail him. He had previously undergone surgery in 1948 to have an abdominal aortic aneurysm reinforced, but that would eventually be the thing that would go on to kill him.

Einstein was largely growing tired of life, although as we have seen he was never growing tired of creating theories or trying to solve problems. However, it did get to the stage where he would pretend that he was ill whenever he simply did not want to take part in something and he also began to appear in less photographs with him using the exact same reason.

There can be little doubt that Einstein did slow down in his last few years, but his mind was still as sharp as ever even up until the day before died. Indeed, he had work sitting with him in hospital, so he was clearly still consumed by a whole host of thoughts until the very last moment.

On the 17th of April 1955, Einstein began to suffer from internal bleeding that was caused by the aneurysm rupturing. He was then moved to hospital where he refused an operation that would have perhaps saved his life with him being quoted as saying.

"I want to go when I want. It is tasteless to prolong life artificially. I have done my share, it is time to go. I will do it elegantly."

Albert Einstein died the following morning in Princeton Hospital. He had previously stated that he wished to be cremated as he appeared to be concerned that people would effectively worship at his bones. He got his wish with his ashes being scattered at a location that was only known to a handful of people.

However, not all of Einstein was cremated.

As Einstein was viewed as being one of the most intelligent human beings to have ever existed, it was no surprise that his brain was then removed for scientific study within eight hours of his death. Since then, it has undergone extensive studies by various institutions around the world and it has certainly led to us developing a better understanding of just what exactly made this individual so different from the majority of us. However, it should be noted that having his brain made available for scientific research was something that Einstein had specifically requested although this is something that has been debated with even his family doubting the validity of the claim.

The Removal of the Brain.

The story of the removal of his brain is one that is actually surrounded by controversy. It was removed by a pathologist based at Princeton University called

Thomas Harvey who then photographed it from various angles and then dissected it into a number of small segments. Pieces of the brain were then sent to various neuroscientists around the world, but Harvey held onto large parts, a fact that remained unknown for some twenty years.

Eventually, parts of Einstein's brain were discovered preserved in to jars in the possession of Harvey with this leading to a media frenzy for a number of days. In defense of Harvey, he was a pathologist and not a neuroscientist and he had sent slivers of the brain to those that understood the brain more than he did and then waited for them to uncover stunning facts and publish it in medical journals.

The problem for Harvey was that, at least initially, there appeared to be nothing too spectacular about the brain of Einstein and no reports appeared anywhere in the world. It seems that Harvey became rather despondent with the lack of progress and it is known that by the 1970's he was living in Kansas and

Einstein's brain was in a jar beside some cider almost as if it had been unceremoniously dumped.

However, that all changed in the 1980s thanks to the work that was being carried out by a scientist called Marian Diamond, but prior to going into her work we should briefly look at the little work that was initially carried out by Harvey.

The first observation by Harvey concluded that the brain of Einstein was actually smaller than average, but it should be noted that the size of the brain is not related to intelligence levels. However, it should also be noted that there were other areas of his brain that were more developed and larger than others and this, in part, contributed to his ability to think in a certain way. This was something that he was unable to observe due to a lack of training on his part, so for decades there were no advancements in the understanding of what made Einstein's brain so special.

As we said, that changed with the work of Diamond which Harvey had become familiar with in the 1980s. He was aware that she was conducting experiments on rats connected to the brain and he contacted her to see if she would be interested in working on some of the samples of Einstein's brain as part of her studies. Clearly this was something she was never going to turn down.

For example, photographs that were taken at the time of the removal of his brain show that two specific areas known as the left angular gyrus and the supermarginal gyrus were enlarged. These areas are not linked to general intelligence, so it did not boost his IQ, but it is known that these areas being enlarged would have had an impact on his mathematical ability as well as something known as visuospatial cognition. To scientists, this explained how he was able to conjure up various equations and understand the concept of time and space that ultimately led to his most famous works.

Further work would be carried out in the future on his brain although it was over an extended period of time.

His Brain and Glial Cells.

It was long assumed that, due to his intelligence, the brain of Einstein must have had more brain cells as surely that would be the only logical explanation for him being able to perform these complex equations. In a certain way this was correct, but not in the way that people initially thought.

Studies have shown that Einstein did have more brain cells, but only of a certain type known as glial cells. For decades scientists had focused primarily on the role of neurons in the brain, so it did require some additional research to be conducted on this type of cell in order to fully understand the difference that it would make to the way in which his brain operated.

These glial cells are often overlooked when, in actual fact, they deserve so much credit for the way in which our brain works. The word itself comes from the Greek word for 'glue' so that gives you some indication as to the role that they play.

These cells are more important than people realized. They play the role of protecting the neurons as well as the cellular networks. Initially it was believed that they simply nourished the cells, but now research has shown that they are connected to the speed at which the different neurons communicate with one another and this, in turn, boosts cognitive ability.

In Einstein's case it was discovered that he had more of these glial cells and in particular in an area of his brain known as the left inferior parietal area. This is important because it is known that this part of the brain is involved in processing information from other areas, so the information was being processed in a shorter period of time.

It is accepted that the more connections between these cells will ultimately lead to there being more communication that is carried out at a faster speed. This increase in connections was also something that was found in the brain of Einstein and, yet again, it was the area of the brain that was important.

It was observed that he had more of these connections in several areas including the temporal lobes, the prefrontal cortex, and the hippocampus. For us, this is important because we are talking about the structures that are responsible for the memory. In other words, Einstein was not only able to think about things quicker than the average human, but his memory in general was also substantially better.

These findings give us a glimpse into the way in which his brain worked, but there is also the question as to why his brain was like this as it throws up the question as to whether he was just born special in some way.

What has been discovered is that rats that were faced with no new challenges or tasks would develop less synaptic connections when compared to those that were forced into thinking about new challenges on a regular basis. In other words, the mere fact that Einstein had been curious about things continually, and from an early age, played a role in the development of his brain. The fact that physics was full of riddles and puzzles with him then continually thinking about them, solving them, and then moving onto the next puzzle led to him developing a greater number of connections.

What we can conclude with his brain and the work of Diamond was that it was different to the average human, but it is the way in which it was different that is important.

However, that is not the entire story of the study of Einstein's brain.

Research by Britt Anderson and Sandra Witelson.

Additional research was then carried out in the 1990s by a scientist called Britt Anderson based out of the University of Alabama. Anderson had focused specifically on certain areas of his brain including his frontal cortex which was apparently thinner than it is in most people. However, this smaller space was packed with more neurons than was normal and this then led to Anderson wishing to explore how this would then have an impact on the way in which Einstein was able to not only think, but make sense of his thoughts.

This led to both Anderson and Harvey contacting Dr Sandra Witelson who was carrying out research at McMaster University in Ontario. Her studies had been looking at the differences between the male and female brain and the discovery of the more densely packed neurons in the brain of Einstein coincided with studies that she herself was conducting. When she was then offered the opportunity to work on a part of Einstein's brain she obviously jumped at the chance.

This in itself was an important step forward in building a better picture of the brain of Einstein. Witelson was helped by the fact that she had a large collection of brains to study covering a range of IQ levels, ages, and even the psychological state of the individual. This meant that she had a far superior control group that could be used in comparison to the brain of Einstein than the work that had been previously carried out by Diamond. In other words, it would be difficult for people to argue against any findings that may occur due to the study.

Harvey himself decided to travel to Canada with his section of Einstein's brain and he allowed Witelson to take almost a fifth of what was left for her study. This was the largest section of his brain to have ever been studied at any given time, but it was hoped that looking at a larger section would allow for more accurate findings to be uncovered.

When faced with Einstein's brain, Witelson decided to focus primarily on the areas known as the parietal and temporal lobes. She also spent a considerable period of time carefully studying the photographs that had been taken on his brain just after his death.

Her Studies.

The first thing that Witelson noticed from the photographs was that a part of the brain known as the Sylvian Fissure was largely missing from his brain. This fissure separates the two different parts of the parietal lobe and as a direct result of this missing it was noticed that the parietal lobe in Einstein's brain was 15% larger than average.

It is amazing to consider that it took some 40 years before this was noticed and its importance cannot be overshadowed. This part of the brain is known to be linked to our mathematical ability, spatial awareness, and also our ability to visualize things in 3D. In other words, this could be attributed to the way in which he was able to conceive of those complex mathematical hypothesis and create those elaborate equations.

According to Withelson, the fact that this fissure was practically missing did mean that the brain cells were largely encouraged to crowd in together allowing more of them to develop and, as a result, there was greater communication between the cells. This allowed his brain to process information faster than the average person.

Furthermore, Withelson hypothesized that this could have contributed to the later development of Einstein when it came to speech. This was something that had been of concern to his parents, but it could have easily been attributed to the way in which the structure of his brain was just different to the norm.

The problem is that Einstein's brain is not actually that remarkable at least not in the way that Harvey perhaps expected it to be. He believed that there would be profound differences, but aside from some subtle changes there is nothing that untoward. This in itself is a surprise, but it does show that there is more to Einstein than just his brain. Yes, he was clearly able to develop certain areas of his brain due to his continual pursuit of solving riddles and puzzles connected to the universe and he was able to process information faster than the norm, but his brain was not in itself superhuman.

There is one other point that must be mentioned. It is accepted that Einstein was a genius, so it has been argued that it is difficult to come to any strong conclusions regarding his brain until we are able to compare it to another genius that is on the same level as Einstein. It is stated by some scientists that we could, potentially, be dealing with a unique brain in some way, but we are unable to know this at this point as our understanding of the brain is still rather limited. Only through further neurological research will we perhaps uncover the truth about what was going on inside the mind of Einstein.

Dealing with his Estate.

Einstein had previously stated that he wished his estate to be bequeathed to the Hebrew University of

Jerusalem and this included the ability to use his image. Considering the way in which he has since appeared on a whole host of items it does appear to be the case that the university will have done extremely well out of this arrangement as they are responsible for licensing where and how it can be used.

The fact he left his estate and the ability to collect royalties to this university hardly comes as a surprise. He was a strong supporter of the university during his life and you may also recall that, at one point he was referred to as being the most important Jew alive. He clearly felt some kind of affinity with the university and the result is that they continue to benefit from his estate to such an extent that they are able to sponsor various science related events as a direct result.

Chapter 11: How Einstein is Viewed.

When we look at Einstein it is important for us to remember that he was purely a theoretical physicist. In other words, he merely sat and worked out theories using nothing more than pen and paper with others then being set the task of either proving or disproving what he said.

However, Einstein is viewed in higher regard than most. He is seen as having had a unique way of being able to look into the workings of the universe in a unique way. He was not merely an abstract thinker, but rather he saw ideas within the universe that were concrete and he then sought to turn them into an explanation that could be understood by everybody.

The remarkable thing about Einstein was that he seemed to just have the ability to identify clear problems with physics and could then set about trying to resolve them. He was able to visualize the different steps that he was required to follow in order to get to an adequate solution. His vision, as well as the fact he saw his own advances as merely being a stepping stone onto something else. He is seen as being rather humble even though he was able to completely change the way in which the universe was viewed.

In order to actually understand the way in which he is viewed it is important that we look at various aspects of his life. Not only do we have to understand the impact he had on the world in general, but also science, and even various aspects of popular culture.

His Impact on the World.

It is important that we look at the impact that Einstein had on the world and for this we have to include not only from a scientific point of view, but also with his other endeavours with a main focus on his role as a pacifist. That in itself is a major topic all on its own, so we will look at it in some detail shortly.

However, we should begin by discussing the impact that he had on the world due to his theories and revelations within the world of physics. There is no doubt that his work between 1905 and 1925 transformed our understanding of so many different aspects of the universe. Indeed, you must remember that he created the idea of cosmology due to his theories and he also brought science and physics into the mainstream. Never before had it really been discussed in any great length by people just going about their daily business, but that is one thing that he did manage to achieve.

His Theories and our World.

The main way in which Einstein has changed the world is via his Theory of General Relativity. Quite simply, without this there are strong arguments that we would not have progressed as we should have as a race.

It could be argued that modern day technology such as GPS systems would simply not exist if it were not for the work carried out by Einstein. This technology focuses on various measurements and the use of satellites and various aspects of the theory put forward by Einstein has to then be taken into account in order for them to be as accurate as they need to be.

It should also be mentioned that the 'Big Bang Theory' is something that has been the focal point of physics for some time. Scientists have often postulated that if we go back far enough in time that the universe would shrink until it got to the point where it all began. However, Einstein's theory on the space and time continuum also suggested that this would have been the case and that the universe is something that is constantly expanding. His theories on the continual expansion have since been proven due to the emergence of more powerful telescopes that are able to reach deeper into the universe than ever before. Their calculations then further support the theories as put forward by Einstein.

Eisntein was also responsible for giving us the idea of black holes in the universe even though he himself doubted their existence. This has led to something that has intrigued people for generations and scientists have now managed to prove that they do indeed exist and that they then have a direct impact on the universe. His Theory of General Relativity was instrumental in our discovery of these black holes which are concentrated areas of gravity whereby nothing is able to escape.

Finally, the idea of traveling at warp speed is something that has often been put forward by science fiction writers as a means of getting around the galaxy. However, scientists have now suggested that this may very well be possible even though our understanding of it is at its earliest stages.

This concept of speed-space travel does stem from Einstein's original theories even though it has taken almost 100 years for this part to have been developed even to the stage where it has been suggested that something is possible. With this, astrophysicists point to the section of his equation whereby it is possible to bend and warp space with this leading to the possibility of traveling at any speed you desire in the universe. The problem is that even though it is possible from a theoretical point of view, we do not possess the technology or materials that would allow us to do it.

It can also be argued that Einstein started the chain of events that ultimately led to the creation of nuclear energy, something that we are all familiar with. The actual development of nuclear power as an energy form was down to the work of other scientists, but it all stems back to his original $E=mc^2$ equation with this being regarded as the real breakthrough in the field. This equation is accepted as having been the stimuli for additional work to be carried out that ultimately led to both nuclear energy and the atomic bomb.

There are a number of other ways in which the science of Einstein has affected our understanding of the world and even our daily lives. However, it is sufficient to say that while other scientists may be linked to certain developments, that they would have struggled to make the same advancements if it was not for this one key theory developed in 1915.

Einstein and Science.

It is easy to understand how Einstein is viewed in the science community and there are various examples of the high regard he was held in during his life. The very fact that scientists some 100 years after the creation of the Theory of General Relativity are still dedicating so much time and energy to his work shows how it has now formed an integral part of so many aspects of physics.

We can also point to him establishing the science of cosmology as playing a key role in the further development and understanding of the universe. Einstein largely became something akin to a shoulder to lean on when it came to certain theories, at least among his closest friends. He would be referenced on a number of occasions and even though some of his leading counterparts did argue against his theories, and disputed them in public forums, he was still highly respected in general.

Einstein the Man.

It is also interesting to look at Einstein the man and how he is viewed in general. As we discuss elsewhere, it is a shame that the most famous photographs that include him are those that portray him as being rather eccentric and clowning around. This just helps to develop the wrong idea of him as a human being and it has fed the theory of the mad scientist.

However, in all honesty he should be viewed in the following manner.

First and foremost he was clearly a genius and there can be no disputing that theory. Furthermore, he should also be viewed as being an individual with a strong conviction of his theories who was not afraid to challenge himself or science even if it meant going against the grain or what was seen as being the norm.

Einstein was also largely a loner when it came to his work. He did collaborate with others, but that was something that reduced in frequency as he got older. Instead, he preferred to lock himself away with his own mind and dedicate himself to his very own theories. This is in direct contrast to the image that is often put forward of him as being a jovial character on a constant basis. The truth of the matter was that there were two very distinct people locked inside one powerful mind.

Einstein and Awards.

We know that Einstein was awarded the Nobel Prize in 1921, but he was also awarded a number of honorary degrees from universities around the world in recognition of his work. These degrees varied from various science related ones, to medicine, and even in the realm of philosophy. Of course these were merely symbolic and were a sign of the gratitude of the university for the work that he had conducted, but it is still an example of how he was held in high regard in the academic world.

However, it was not just universities that bestowed him with honors as it is difficult to find a scientific academy in the world that did not provide him with some kind of recognition. This was matched by being given awards from the scientific community including the Franklin Medal from the Franklin Institute in 1935 and before that the Copley Medal in 1925 from the Royal Society of London.

You must also remember that he was invited to give lectures around the world due to his advances in the world of physics. This was a time where travel was not as easy as it is now, but he still found his way around the world with lectures across Europe, North America, and even the Far East. No other scientist had ever had to do this in response to their work so it is just another example of how he was viewed in such high regard that people even sent out invites for lectures in the first place.

Einstein and Politics.

Einstein was also known to have a keen interest in politics. However, his main focus was on issues of peace as well as social justice rather than becoming involved in other causes. This was clearly something that troubled him from an early age as can be seen in the way that he rebelled against the type of education that he was forced to enter as a child. He is known to have had issues with the concept of German education being authoritative and being taught with something akin to military precision, so he was never afraid to stand up against what was accepted as being the norm.

We can also see that World War I played a role in shaping his views particularly on world peace and being a pacifist. You must remember that the war was extremely bloody with a vast number of men being killed in battle whereby entire towns and villages would have the male population effective being wiped out. To Einstein, this was not the way that international politics should be conducted. To him, resolving issues without fighting should be paramount, so even if he was not Jewish we already know that he would have refused to fight for the Nazi Party in World War 2.

Einstein and Being a Pacifist.

It is clear that Einstein felt that his new-found fame could be put to good use in the political world during the 1920's. He was loved by the press, and was also known to the public which was a rare thing for a scientist, with him holding the belief that he could perhaps make a difference.

This led to him becoming a strong supporter of the anti-war movement in the 1920's and he also fully supported the concept of the conscientious objector. This part regarding providing people with the option of not fighting was certainly regarded as being revolutionary as conscription had played a major role in the war where not fighting was often seen as being a sign of weakness.

Einstein had been a pacifist for his entire life. Indeed, he had referred to himself as being a militant pacifist as he believed in fighting for peace, which he realized went against the entire idea of encouraging more development in the realm of nuclear power. We can even look at this position he took in 1914 at the outbreak of World War I whereby most scientists and academics in general were producing articles that were merely attempts to justify the invasion of Belgium by Germany. This was done to appease the Kaiser, but Einstein made it clear that he was against such an event even though it would potentially alienate him from society in general.

Instead, we know that Einstein instead produced what can only be described as an anti-war manifesto. In this manifesto he put forward his plans for cooperation between not the countries that were involved in the war, but cooperation between scholars. His view was that Europe would need these scholars to work together in order to effectively rebuild itself once the war was over. He then wanted to take things even further by creating what he called a 'League of Europeans'.

Even though this approach did anger a number of people, it did not stop Einstein from continuing to pursue this idea even throughout the duration of the war. He also remained in Berlin and made repeated attempts at trying to create peace even though the odds were firmly stacked against him. As a result, we know that he then buried himself in his work in a blatant attempt to largely block out what he saw as the agony of Europe that was going on around him. This must have been a difficult time for such a staunch pacifist although it can be argued that burying himself into his work eventually led to the Theory of Relativity as we know that he focused on this during these years.

However, he did continually campaign for a democratic government to take control of Germany even during the war years. He was not afraid to write letters to various people in positions of power and he had a tendency to ask some rather searching questions that were certainly regarded as being out of the norm.

This desire for a democratic government was one that Einstein then looked back on with some sadness. He is quoted several years later in a letter to a friend staying.

"Do you remember when we took a trolley-car to the Reichstag, convinced we could turn those fellows into honest democrats? How naive we were even at the age of 40! It makes me laugh to think of it."

His position as a pacifist did have some impact on his career in Germany. After the end of World War I, Einstein was in great demand and this led to him meeting famous figures such as Winston Churchill, President Harding and even Charlie Chaplin. However, in his home land this was not the case. A number of Germans were seeking some kind of an explanation as to why Germany had lost the war and they came to two conclusions as to who was to blame.

Jews and pacifists. This was a problem for Einstein as he was both, so the tide did turn against him to a certain extent.

This was not something that was going to put him off being a pacifist. After some time, and due to his continuing rise in fame, Einstein was asked on a number of occasions to talk to the German Parliament. To him, this was an opportunity to further his pacifist cause and he was quick to discuss the merits of non-violence as well as the rise of nationalism. Instead, he effectively pleaded for some form of international cooperation and for there to be some kind of international authority, which does come across as a pre-cursor for the United Nations.

The problem that Einstein faced regarding his approach to violence and war was trying to come to terms with the extent to which scientists had to be held liable for their discoveries. Just exactly how responsible are they for what then happens with their work even long after they have moved on in their life?

This was certainly something that would bother Einstein particularly during the 1930s.

However, we should also mention a journey Einstein took in 1922 through Northern France that really managed to reinforce his pacifist ideas. During the journey, he found himself in the area of the main battles of the war and encountered the destruction that was still all too apparent even four years after the war had ended. It was at this point that he was quoted as saying '*War is a terrible thing, and must be abolished at all costs.*' This was one saying that he would repeat a number of times during his life and, in a way, it became his mantra in connection to his desire for world peace.

His desire to appear at the head of the pacifist and anti-war movement was certainly something that was important to him even when it was clear that his life was in danger. In the summer of 1922, a prominent member of the German government was assassinated due to his stance regarding peace and pacifism, and he was also Jewish although this perhaps did not play an important role, and Einstein knew that he was also at risk. However, he refused to bow down to these pressures and instead he continued to appear at various rallies and demonstrations linked to the anti-war movement directly going against the wishes of those closest to him. It is clear that his old teacher who commented that Einstein never listened to anybody else was more accurate in his observation than her perhaps realized.

By 1923, Einstein had been invited to visit Japan, a country that he then fell in love with due to the way in which it appeared to push forward an idea of peace. However, he was in for a rude awakening by the time he returned to Europe as fascism was on the increase across the continent with the likes of Mussolini coming to power in Italy.

If you recall, Einstein had become a member of the supposedly powerful League of Nations Committee on Intellectual Co-operation and even though he was happy to join he certainly became rather frustrated at their lack of ability to actually do anything. This was put to the test in 1923 after France invaded part of Germany as a way of trying to recoup damages for the war. In particular, they invaded the Ruhr Valley and there was a real sense of disappointment from Einstein in that they were unable to help push forward a suitable conclusion to the situation.

As a direct result, Einstein resigned from the committee stating that:

"I have become convinced that the League has neither the strength nor the sincere desire it needs to achieve its aims. As a convinced pacifist, I request that you strike my name from your list of members. By its silence and its actions, the League functions as a tool of those nations which, at this point of history, happen to be the dominant powers."

In other words, he felt that the League was actually useless with what it was wanting to achieve since it simply did not have the power to make any relevant changes.

However, Einstein did have a change of heart just a year later when he asked to rejoin the League. At this point he summed up his feelings by saying:

"I've come to feel that I was influenced more by a mood of disillusionment than by clear thinking."

The other members of the League were only too happy to, once again, have Einstein on board. No grudges were held against him and indeed it is known that he also serenaded them at one point with his violin, which would have certainly been a rather bizarre event to witness.

This remained the status quo for Einstein with the league until 1930. Up until that point he had been a regular attendee at meetings, but once again he became disappointed in the way it operated. For some, part of the problem was that Einstein was primarily used to working all on his own, so the mere thought of working as a group or discussing things in meetings was just so alien that he was unwilling to participate any longer.

Einstein and Resisting War.

Einstein, as we know, was a strong advocate for peace over war and his own opinions coincided with an increase in the number of people that were directly resistant to war after 1914. By 1928, Einstein was certainly becoming far more vocal than ever before regarding his push for peace and in that year he signed an international manifesto that stated it was against the concept of military conscription.

Also in 1928, Einstein was asked to join the board of the German League for Human Rights, which was a pacifist organization. He was only too happy to do so leading to the creation of a special speech to commemorate the 10th anniversary of the Armistice. In his speech, he voiced his concern about humanity and clearly had concerns that the 1914-1918 war would not be the only one of its kind.

"The political apathy of people in time of peace indicates that they will readily let themselves be led to slaughter later. Because today they lack even the courage to sign their names in support of disarmament, they will be compelled to shed their blood tomorrow."

This was certainly a powerful statement to make, but Einstein was aware that he had real influence with the power of his words, so his speech was certainly something he would have poured over for some time. He then took this concept even further the following year in an interview in a Czech journal where he made his thoughts on conscription even clearer.

"I would unconditionally refuse all war service, direct or indirect, and would seek to persuade my friends to take up the same stance, regardless of how I felt about the causes of any particular war."

Einstein did also make his famous 2% speech in the United States in 1930, which tells us how this was a concept that remained dear to his heart. Indeed, whereas before he was often talking more to other leading figures, it does seem that this speech was aimed more at the normal citizen who would, of course, have been directly affected by the idea of conscription.

"In countries where conscription exists, the true pacifist must refuse military duty. In countries where where compulsory military service does not exist, true pacifists must publicly declare that they will not take up arms in any circumstances....The timid may say, "What's the use? I will be sent to prison." To them I say: even if only two per cent announced their refusal to fight, governments would be powerless – they would not dare to send such a huge number to prison."

Einstein and the Concept of Disarmament.

We already understand how Einstein was a strong supporter of the anti-war movement, but he was also a strong campaigner for the concept of disarmament. To him, this was something that just had to happen if the world was to avoid another war that could ultimately destroy humanity.

However, this was something that he preferred to do from behind his own desk and with as little interference from others as possible. To Einstein, it was important that he was able to get his point of view across to as many influential people as possible. This led to the writing of numerous letters to heads of state around the world or heads of government departments. The Finnish government received letters, individuals in Bulgaria were contacted by Einstein, even Polish courts were sent letters when they were involved in trials connected to human rights issues. Even though he was a Zionist, Einstein was even concerned with the rights of the Palestinians and was not afraid to voice his opinion in that area either even though it went against so many influential Jews.

Closer to home, Einstein began to oppose the militia system that was in operation in Switzerland. With this, we even have quotes from him to those involved in the pacifist movement even when pressure was being applied by the government. *"Let me express my respect for your courage and integrity."* he said, *"A man like yourself acts like a grain of sand in a machine; by such grains the war machine will be destroyed."*

Furthermore, Einstein also sent a letter of support to Belgians who were on a peace pilgrimage. He donated money to a war resistance movement that was based in Denmark. He also contributed to the raising of funds to establish the group called War Resisters International. Clearly, the process of disarmament was one that he took very seriously indeed and he dedicated a substantial amount of his time to this cause.

To many, Einstein was a driving force in this movement. When the WRI held its conference in France in 1931, Einstein sent a message of support that really managed to hit home at the absolute core of the issue.

"You may become the most effective group of men and women involved in the greatest of human endeavours. The people of 56 countries whom you represent have a potential power far mightier than the sword. All the nations of the world are talking about disarmament. You must teach them to do more than just talk. The people must take it out of the hands of statesmen and diplomats. Only they themselves can bring disarmament into the world."

It must be pointed out that this was not purely Einstein on some kind of ego trip as it was a real movement that gathered pace. A number of countries around the world actively sought disarmament resulting in a meeting being held in Geneva in 1932 to discuss this very idea. However, political debate got in the way of an actual resolution and such was his frustration at the lack of progress that Einstein even visited Geneva and held a press conference. At it, he voiced his displeasure at what was going on.

"If the implications weren't so tragic, the Conference's methods would only be called absurd.

One doesn't make wars less likely to happen by formulating rules of warfare...The solution to the peace problem can't be left in the hands of governments...I think the conference is heading for a bad compromise. Whatever agreement is made about the "types of arms permissible in war" would be broken as soon as war began. War can't be humanised. It can only be abolished."

These were indeed stern words and it is obvious he is calling into question the entire approach of the conference. To him, it was just ludicrous for governments to attempt to establish rules of what was and was not allowed in war because, in difficult times, nobody would then adhere to them. There was only one solution that was capable of working, complete disarmament.

Pacifism in the 30's.

Even though Einstein moved to the United States in the early 1930's, he still pursued his thoughts on pacifism. Prior to leaving for the US he did agree to put his name to a manifesto that was backed by the Womans International League for Peace and Freedom which focused on the concept of world disarmament, a cause that he was only too keen to support.

During one of his trips to the United States in 1930, he then gave a speech that was designed to encourage people to give up the concept of war and fighting for their country. You must remember that he was a staunch supporter of the concept of the conscientious objector which led to him stating that is only two percent of those individuals called up to fight for their country refused to do so, then governments and the army would become completely powerless in an instant. His argument was that it would be impossible, and impractical, for them to go ahead and imprison so many people. To some, this was mutiny, but to Einstein it was merely the individual being able to make a decision as to whether or not they wished to support military action.

This also meant that Einstein had strong opinions on the concept of conscription into the army. He was strongly against such an idea, as can be seen by the way he avoided it himself by changing nationality, and he sought support in the international community for those that were conscientious objectors to be protected for their views rather than being punished.

To Einstein, conscription was similar to the individual becoming enslaved by their government. He felt that the only way in which peace could ever come around, and provide genuine security to society, was by the process of disarmament. Without that, there would always be that risk of war and individuals being forced into taking part even against their wishes.

Seeking Support for the Pacifist Movement.

We know that the concept of pacifism was clearly something that Einstein held dear. Throughout the course of the 1930's he was responsible for a number of articles being produced that focused on the concept of peace in different formats as well as the production of various letters to prominent individuals seeking their feedback on his stance.

To Einstein the idea of peace was easy to understand. To him, it should have been obvious that humanity had to be more important than the aims and intentions of individual nations. However, he also foresaw a problem in that it was pointless waiting for leaders of nations to give up their ideas surrounding war as he believed that this would be difficult to achieve. Instead, change had to come from the people as numbers talk and that would ultimately lead to nations being forced into changing their approach. In his mind, it was puzzling how people that claimed to be upstanding members of humanity could engage in war when they knew that so many innocent people would be killed. This was a tragedy beyond any description and it was an approach that even the great mind of Einstein was unable to comprehend.

Einstein and Freud.

When we move into 1931, Einstein was invited by the International Institute for Intellectual Cooperation to enter into a discussion with a prominent individual of his choosing regarding a matter that was important to society and humanity. Einstein was only too happy to accept this invitation and he elected to enter into a discussion with Sigmund Freud. His subject? Any suggestions on how man would ever be able to break away from the concept of war.

In a rather extensive letter to Freud, Einstein set out what he saw as being a problem in that advances in technology were ultimately leading to greater problems with war and that everyday people would suffer as a result. He commented on issues surrounding nationalism and people being biased towards it and partly attributed this concept to the problem of the resurgence of war and wished to know how this could be tackled. Einstein could see that the world was going down a particular path and it was one that he wished to have no part in.

To Einstein, this was his opportunity to get some answers to aspects of humanity that puzzled him and who better to ask than the eminent psychotherapist,

Freud. He believed that nobody was better placed to resolve the issues that he had surrounding humanity.

To his credit, Freud did become involved in the debate with a lengthy and well thought out reply whereby he put forward his ideas as to why humanity appears to prefer to act in certain ways. To him, the idea of conflict is just something that was part of human nature with it directly related to our emotions. However, he also stated how rational behavior can help to overcome those emotions so, in his eyes, the only hope for humanity to avoid war was if we somehow managed to become more rational in our way of thinking.

Going Back to 1932.

If we go back to 1932, which we discussed earlier regarding a convention in Geneva, then this really was a difficult year for Einstein and the pacifist movement.

It had previously been arranged for the creation of an International Peace Center that would be based at The Hague in the Netherlands, but that never materialized. The reason was that a number of international pacifist groups refused to take part and in the process a lot of its potential power was lost. Furthermore, it was also the year when an anti-war congress was arranged in Amsterdam. This did go ahead, but unfortunately for the congress it was hijacked by the USSR which sought to score points from a political perspective and to push forward Communist ideology. As a result of this, Einstein even found himself being referred to as a Communist, a claim that was not only incorrect, but also one that caused him several issues on a personal level.

Einstein and Dealing with Nazism.

The Rise Of Nazism and Problems for Einstein.

The 1930's were a difficult time for Einstein in that he was working in Berlin at the same time as Hitler was coming into power. Einstein was a known pacifist and he also had one other problem, he was Jewish.

Both of these things would clearly put him in direct conflict with the ideas of the Nazi Party and Einstein certainly had no inclination for putting up any kind of a challenge.

The fact that he was Jewish, and thanks to the views that were held by the Nazi Party, it led to other physicists and scientists in Germany eventually tagging his work as being Jewish physics and attempting to discredit some of his theories. Ultimately, they clearly felt compelled to portray his work as some kind of Jewish conspiracy, which does fit in with how the Nazi Party viewed Jews in general. Einstein was also not the only prominent scientist to leave Germany as many others followed due to them holding a serious belief that Hitler would attempt to create an atomic bomb and they wanted to play no part in it whatsoever.

The problem that the Nazi Party had was that Einstein had already been regarded for a number of years as being a genius and this was something that they could not allow to happen. Instead, they were forced to produce propaganda whereby he was portrayed more as a Jew with rather limited abilities, therefore reducing his impact in the scientific world.

It is interesting to note that this propaganda extended to years after he left Germany with the Nazi Party incorrectly stating that he was acting like some kind of messiah to the Jews. They also stated that he was nothing more than a tourist attraction in the United States and that his theories had been forgotten about and were only mentioned when Einstein himself popped up and mentioned them. Reports from the time, that were written by the Nazi Party, refer to Relativity as being nothing more than a huge bluff on his part and they then accused him of discrediting various other individuals going back several decades. In other words, there were continual blatant attempts at destroying his reputation although, thankfully, this was something that was pretty clear to everybody.

However, that was not the only problem that Einstein was facing at this time.

The issues faced by Jews across Germany even extended to the point whereby they were banned from taking part in any university work. We also know that Einstein was put on a hit list to be killed as he was simply regarded as being a prominent Jew who was being held in high regard. Clearly, this was not something that could be allowed to continue.

Indeed, when you consider the fact that Einstein was doing relatively well for himself due to his touring and being paid to speak at various institutions, while others in Germany were falling on hard times, it led to some hostility towards him from everyday people.

We previously stated how Einstein fled Germany at the start of the rise of Nazism in his home country with him leaving for the United States and the job at Princeton University. This was not Einstein being paranoid as it was clear that he would be singled out for 'special' treatment by the Nazi Party due to his prominent position as well as being Jewish. After leaving for the United States on a tour, it was actually not too long until his home in Germany was seized by the authorities.

This move, while perhaps not entirely unexpected, still took Einstein by surprise and he is known to have spent time in Belgium trying to take stock of the changing situation.

The interesting point to examine here is that his feelings regarding disarmament and war resistance did change slightly thanks to the rise of the Nazi Party. When we move into 1933, Einstein believed that all that would be required would be economic pressure placed on the Nazi Party for things to change. However, by the time we move into the summer of 1933, his attitude had changed as he felt that an international peace force would be required to take control. Finally, this approach evolved even further when he gave his support to the concept that was put forward by the New Commonwealth Society. This concept was that there had to be an international army put in place to deal with events such as this, so even though he was in favor of disarmament, he did begin to realize that perhaps it was not the appropriate solution in every instance.

This idea of a change in his stance with resisting war can be seen in his exchanges with individuals in Belgium, a country to which he had close ties. Indeed, he informed the King of Belgium that, in his eyes, any conscientious objector should be given the opportunity to take part in some form of alternative service to help the war rather than being directly involved in the fighting. Furthermore, he was quoted as saying to a friend who was also against military action that his opinion was:

"If I were Belgian I would not, in the present situation, refuse military service. I would enter it in the belief that I was helping European civilization."

What we can infer from this is that Einstein was aware that there was a bigger threat to world peace than what could be contained by a completely pacifist approach. However, this apparent change in beliefs did lead to other pacifists ridiculing Einstein and reacting in horror at this approach. Einstein, to his credit, stood firm in his beliefs and even tried to clarify his stance in order to preserve the fact that he was still a pacifist.

"I loathe all armies and any kind of violence; yet I'm firmly convinced that at present these hateful weapons offer the only effective protection. Should Nazi militarism prevail you can be sure that the last remnants of personal freedom in Europe will be destroyed."

Even though he made enemies within the pacifist movement, there can be little doubt that it was still one of Einstein's core beliefs. He had personal experience of how things were changing in Germany and he was only too aware that the country was not only re-arming itself, but it was doing so at an alarming rate.

He had become rather despondent at the way in which the world had been unable to come to an agreement as to how they could deal with this kind of threat in a peaceful way. It seemed even to Einstein that governments around the world had decided that you could only fight fire with fire, an idea that went against everything he believed in.

Einstein and the Push for International Peace with World War II.

Even after effectively being forced into exile in the United States in 1933, Einstein remained focused on the idea of international peace and still held onto the dream that there could be a united, concerted effort involving key individuals around the world.

It was clear that tension was rising in Europe due to the rise of the Nazi Party, but Einstein held onto his belief that war would eventually become outlawed as the people would simply state they had experienced enough war once and for all.

By 1938, the picture in Germany was looking bleaker than ever. Hitler had come to power and the persecution of the Jews had already started. Einstein looked on in horror at what was happening in his homeland and he asked non-Jews in both America and Europe to work together to stop this dreadful situation. He knew that things would only get worse and sought to do anything he could in his power to prevent things from deteriorating any further.

"No government has the right to conduct a systematic campaign of physical destruction of any segment of the population which resides within its borders. Germany has embarked on such a path in its inhuman persecution of German and Austrian Jews...Can there be anything more humiliating for our generation than to feel compelled to request that innocent people not be killed?"

This was a fight that Einstein would eventually lose although it is clear that he largely foresaw the problems that would befall the Jewish population in Nazi held territory.

Einstein and Views on the Responsibility of Scientists.

Another key area to be examined is the way in which Einstein viewed the level of responsibility that scientists should have surrounding their experiments and discoveries.

The main area of concern was certainly surrounding atomic power. The atom was discovered in 1922 by Rutherford and at the time there was certainly speculation as to how this new discovery would change the world. People wondered about what would happen with this new form of power although Rutherford himself was quick to point out that this was not the reason why he sought to explain the existence of the atom.

Einstein certainly backed Rutherford on this claim as he had always held the belief that scientists had to be given free reign, to a certain extent, to make new discoveries and explore whatever was possible. Indeed, on this matter he was quoted as saying,

"We must not condemn man because his inventiveness and patient conquest of the forces of nature are exploited for false and destructive purposes. The line of demarcation doesn't lie between scientists and non-scientists; it lies between responsible, honest people, and the others. In our time, scientists and engineers carry a particularly heavy burden of moral responsibility, because the development of military means of mass destruction is dependent on their work."

Einstein was aware that scientists could only really be held responsible for their discoveries and if they were made due to an actual desire to uncover something that was beneficial to the understanding of the universe, then that was something for the good of humanity. However, he understood that this did rely on people that were honest and had good intentions, something that you could certainly never guarantee.

To Einstein, there was a clear difference between scientific discoveries and the way in which they were being used. In his opinion, the only way to be sure that scientific discoveries would not be used in an incorrect way was by making sure that the people elected the correct leaders that had no such desire to pursue science in this manner.

That was one reason why he took his job as a scientist and pacifist so seriously. He wanted to show people that war was indeed a horrific thing to happen, which they already knew, but that power was in their hands. Additionally, he had rather strong views that science itself should never be closely involved with either the military or the state. Instead, it should be capable of working on its own and only subjected to civilian control rather than it being ordered to follow certain lines that would only serve to further the cause of the state or as part of the war machine.

Also, he was only too aware that scientists did have to adopt a moral stance on their discoveries and had a responsibility to be completely open about what could potentially happen with what they had unearthed or created. He felt that the civilian population had to understand that there was a possibility that their breakthroughs could be used in various ways, potentially both good and bad. There had to be a sense of clarity in all of this rather than the civilian population having a surprise sprung on them in the future, at which point it would often be too late to change the course of events.

For this, scientists had to take responsibility and to be aware that their discoveries could, in effect, be hijacked for something that was never their intention. This in itself was a frightening thought for Einstein, but at the same time he had always kept to the idea that science itself could not be restricted in any way and that it had to be allowed to evolve naturally or wherever the mind of the scientist would take it.

He felt that science that was used for other motives was more because of the way in which mankind would try any means necessary to gain the upper hand in times of war or conquest. At that point he then referred to science as being as dangerous as giving a razor to a child to stress that it became a tool that could be abused but that it was important to then view it in a different light from what its original intention may have been. Ultimately, whether or not science was used in a negative and dangerous way was down more to the sense of morality of key individuals rather than the scientist that perhaps made the initial discovery.

Einstein and Popular Culture.

One fact that is rather interesting about Einstein is the way in which his name has been incorporated into common culture. You only have to look at how we refer to anybody that is seen as being some kind of genius by his very name to see how he is inextricably linked to the entire concept of simply being more intelligent than others. However, in a strange twist, his name is also used to describe the opposite, which is something that would have surely appealed to him and his sense of humor. When you think about it, his name is used in a sarcastic manner when an individual states the obvious, which is as far away from being a genius as is humanly possible.

His fame and use in popular culture is something that is still going on even some 60 years after his death. There can be little doubt that he is the most famous scientist to have ever lived and in various polls he always comes out top when people are asked to name people from this field. Indeed, in a number of instances he is the only individual that people can name.

We can also point to Time magazine as a sign of how popular and important a figure he was in his life since they voted him as the most popular individual to have lived in the 20th century. This in itself is pretty astonishing when you consider some of the other individuals that were alive around the same time, but it does show the impact that he was able to have.

You must remember that he appeared on the front pages of newspapers around the world after the revealing of his Theory of General Relativity and his lectures were not just limited to other scientists. Instead, he was being courted by different heads of state and even celebrities from the time were getting in on the act. It seemed that everybody was wanting a piece of Einstein.

Einstein as an Iconic Figure.

It is seen by many as being rather unfortunate that the most popular image of Einstein is the famous photograph of him with his tongue sticking out. However, he was known to be a bit of a joker, so this was not exactly out of character for him, but it does mean that the serious side to his life and work is slightly hidden by this different demeanour.

Unfortunately, it has also led to people linking him to the concept of being a 'mad scientist' whereas he was an actual genius with the two things being distinctly different from one another.

The strange thing about this image is that it has been reproduced on numerous occasions and it has appeared on items ranging from t-shirts, to mugs, and even on mouse mats. The story behind the photograph is that it was taken after he was leaving a party to celebrate his 72nd birthday. Such was the clamour for him that the press were waiting outside hoping to get a snapshot of the world famous Einstein and they are known to have been pressuring him into smiling. However, Einstein had been doing this all day and by the time it got to the evening he had grown tired of the fake smile. As a result, he stuck out his tongue leading to the image that so many people love of him.

This is not the only photograph of Einstein that people are aware of, but the interesting part from our perspective is that the three or four images that are used over and over again never show his work. Instead, we see him riding on a bicycle or even taking part in some artistic style shot where he looks rather longingly out of the picture.

However, it is also important to point out that the most famous photographs of him are all taken from the later years of his life. This is, in part, due to him having a rather more stern looking face when he was younger and he was certainly not as photogenic.

Einstein and Influencing Painting and Art.

The idea of Einstein being able to influence art may sound rather bizarre to some, but this is actually something that happened, especially with Salvador Dali.

It is known that Dali had become interested in the entire concept of relativity and this has been linked to his painting 'The Persistence of Memory' whereby the watch that features in it is melting with this being linked to the idea of time and space distortion straight from Einstein's theory.

Einstein and Literature.

Even though the idea of Einstein influencing art may sound strange, it is less peculiar to think of him being able to influence literature and in particular the realm of Science Fiction.

The list of books that include references to his work is too extensive to list here, but there can be little doubt that his links to the time-space theory as well as black holes or wormholes remain the most widely used aspect of his theories within this kind of literature. As was mentioned elsewhere, it is difficult for us to even think about the likes of Star Trek existing without the work of Einstein.

The interesting part for us is that it could be argued there are two different versions of Einstein when it comes to popular culture. For some, there is the genius scientist who is capable of producing out of this world theories that completely change the way in which we view so many aspects of the universe. On the other hand, you have an Einstein who is full of pretty ordinary quirks and mannerisms that show a completely different side.

Einstein and the Movies.

Einstein has also been portrayed in a number of movies over the year even to the extent of him being the influence behind the character of Yoda in Star Wars. He has also appeared as a holograph in movies, different actors have played him or even characters that are strongly based on him, and it can be argued that his theories have also been used even to create entire plots for a movie. In other words, both his work and his own self have bridged a gap between science and Hollywood that, up until now, nobody even knew existed.

His Work and Popular Culture.

His theories have also managed to partly slide over into popular culture as can be seen by the way in which the majority of people in the world will be only too aware of $E=mc^2$. Indeed, this equation has since entered culture as a symbol for indicating that very intelligent people are hard at work on something that is quite complex. The rather ironic thing is that people can quote the equation, but then they do not actually understand what it means although, in a strange way, this only adds to the power that it appears to hold over the population.

To some, there is a theory that the man that is identified with pop culture has actually now surpassed the real man behind it all. This is, of course, due to his popularity after this famous theory and we can even point to a piece in 'The New York Times' whereby they relay a story attributed to Einstein that gives us a glimpse into another aspect of his life.

He was clearly taken aback by the way in which everyday people had taken an interest in his theory of General Relativity and he is known to have adopted a particular stance when confronted by members of the public. His main complaint was the way in which they would stop him in the street in the United States and simply ask him to explain his theory. This became rather tiresome after a while resulting in him pretending that he was not Albert Einstein. He would voice his displeasure at, yet again, being mistaken for him when all that was actually going on was that he resembled his appearance.

If we think of the mad scientist idea, then it is Einstein that pops into our head, but he is partly to blame for that and not just because of that photograph with his tongue sticking out. He had wild hair, he was rather unkempt looking and it is no coincidence that if we are asked to describe how a mad scientist looks that we then come up with something along these lines. It is even crazy to now think that his brain has been made into an app and this would have surely confused even this greatest of minds.

Chapter 12: The Overall Conclusion of Einstein.

Considering he came from relatively humble beginnings, it is astonishing to think that Einstein then became such a prominent figure in the world in general. It is clear that his encounter with a compass at the age of five as well as the work of his father in electricity played a major role in how he developed as a child.

Albert Einstein went from being a late developer to the most widely acclaimed genius that the world has ever seen. He conjured up startling new theories that changed the world of physics as well as the way in which we viewed the universe. He made the leap from being a scientist to a world famous celebrity who was courted by politicians, leading peace figures, and even celebrities who wished to have a peace of this rather eccentric scientist. Never before had a scientist had this kind of impact across various industries. The fact that actors wished to dine with him is astounding.

The impact that he has had on the world cannot be underestimated even if you are not the type of individual that has much of an interest in the world of science. He has entered into our popular culture with his very name being used as a direct replacement for referring to an individual as a genius. His image is well known around the globe even if his most famous photograph does not accurately tell the story of Einstein.

There can be little doubting the impact he had in the world of science. He made such huge leaps in our understanding of so many aspects of physics that they were the biggest developments perhaps ever seen in the world, or at least since the time of Newton, and even today physicists continue to work and develop his theories. It is easy to see how his name will be up there with the likes of Newton even centuries later.

However, as we have shown there was another side to Einstein, a side that was perhaps more important to the general public who simply did not grasp the magnitude of his scientific discoveries. That side was the way in which he pursued the concept of peace and that is why we spent such an extensive period of time discussing the different avenues he explored in the name of pacifism.

To see Einstein as a mad scientist is certainly wrong. Instead, it is important that people look beyond the photographs of him riding his bicycle or sticking his tongue out and see what he really stood for and what he achieved. Yes he had unkempt hair that blew around and it did look slightly manic in style, but this was undoubtedly a man that was a genius on so many levels and had a tendency to attempt to take the moral stance on every occasion. With this, some would argue that it was difficult to do considering he did have a tendency to have a roving eye.

However, it could perhaps be argued that Einstein was only too aware of the concept of the mad scientist and he was certainly enough of a joker to then play along as some kind of game in his own mind. The fact that his work goes against that theory or image is certainly one thing that cannot be ignored.

He was not afraid to step on the toes of those that perhaps viewed themselves as being more powerful than him if he believed in his cause. How else could you explain him trying to find some common ground with the Soviet Union while living in the United States just to try to avert the Cold War? How could a scientist ever believe that they had the power to be able to do this when they were best known for some equations rather than trying to bring about world peace?

Einstein battled against being supremely intelligent from an early age as well as dealing with a stubborn streak that did mean he was his own man even as a child. He had a self-confidence about himself even when dealing with the most complex of calculations and theories, but at the same time he was always aware of the way in which his theories would be viewed by the scientific community. This was something that did trouble him, but he was wise enough to understand that it happened as he too was guilty of doing the exact same thing to other theories. He was not afraid of getting into scientific arguments as long as it was for the greater good of our understanding of how the world in general operated.

The fame he achieved certainly started him, but he was then intelligent enough to attempt to put it to good use and in his case it was for the pacifist agenda. For that, he should be applauded considering he was attempting to do this at a time when war was high on the agenda. His list of achievements in the world of pacifism and trying to stop rearmament is something that he should have been applauded for at the time although hindsight has perhaps taught us more about this part of Einstein than what was previously known. He clearly never tried to hide from what he saw as being his duty to use his influence and power and to put it to good use and he did get close to disarmament in the 1930's.

The unfortunate thing is that people are generally unaware of this pacifism as they focus primarily on his work in the field of science. This in itself would be a disappointment to Einstein who put real value on his attempts to stop war and to even stop the battle between both the United States and the Soviet Union. Indeed, it could be argued that if politicians of the time had taken more notice of what he was doing, then the history of the world could have been very different from how things then panned out.

Albert Einstein has already gone down in history as one of the most influential scientists of all time and he is up there with the other 'greats' that will be discussed for centuries to come in schools around the world. The impact that he had on the world of physics is still being felt even today with modern breakthroughs being directly attributed to work that he carried out a century ago. He changed our understanding of so much and he also somehow managed to make science seem so much more fun than ever before.

Einstein broke the mould when it came to a scientist. He removed the concept of them being aloof or too intelligent for normal people to engage with them. He was friendly, open to discussions, but he still had an eccentric side that would come out from time to time. It is still astonishing to think that even some 60 years after his death that his image is still being used in various formats. However, it is fair to say that this was a complete accident and it may have been different if he had never put out his tongue.

The one thing that is certain is that our understanding of the world would be completely different if it was not for Albert Einstein and for that we should all be eternally grateful. You may not fully understand his various theories and who could blame you, but it is enough to accept that this is one person that we should be glad was alive.

Made in the USA
Las Vegas, NV
09 April 2024

88473649R00115